# VENTURE JAPAN

### HOW GROWING COMPANIES WORLDWIDE CAN TAP INTO THE JAPANESE VENTURE CAPITAL MARKETS

## JAMES W. BORTON

**PROBUS PUBLISHING COMPANY**
**Chicago, Illinois**
**Cambridge, England**

© 1992, James W. Borton

ALL RIGHTS RESERVED. No part of this publication may be reproduced, stored in a retrieval system, or transmitted by any means, electronic, mechanical, photocopying, recording, or otherwise, without the prior written permission of the publisher and the copyright holder.

This publication is designed to provide accurate and authoritative information in regard to the subject matter covered. It is sold with the understanding that neither the author nor the publisher are engaged in rendering legal, accounting or other professional service. If legal or other expert assistance is required, the services of a competent professional should be sought.

ISBN 1-55738-266-2

Printed in the United States of America

EB

1  2  3  4  5  6  7  8  9  0

Dedication

For my father, my late mother, Travis and Isabel

# Contents

| | | |
|---|---|---|
| Acknowledgements | | vii |
| Introduction | Japanese Venture Capital: A Historical Overview | ix |
| Chapter One | The Internationalization of Venture Capital | 1 |
| Chapter Two | Japanese Venture Capital in the United States | 13 |
| Chapter Three | The Euro-Japan Finance Network | 25 |
| Chapter Four | The Promotion of Venture Business and the Venture Capital Industry | 31 |
| Chapter Five | International Business Strategies of Nippon Enterprise Development | 61 |
| Chapter Six | Japan as a Source of Capital and Debt Financing for U.S. High-Tech Companies | 71 |
| Chapter Seven | The Over-the-Counter Market in Japan | 85 |
| Chapter Eight | Benefitting from Japanese Corporate Venture Capital: Opportunities and Challenges | 113 |

## Contents

| | | |
|---|---|---|
| Chapter Nine | U.S.-Japan Strategic Partnership: The Use of Technology Transfer and International Network | 121 |
| Chapter Ten | Entrepreneurs in Japan and Silicon Valley: A Study of Perceived Differences | 137 |
| Chapter Eleven | Accessing Foreign Venture Capital Sources | 153 |

| | |
|---|---|
| Japanese Venture Capital Sources | 159 |
| International Venture Capital Associations | 211 |
| Index | 219 |

# Acknowledgements

This book belongs to many people. In fact, this collection of articles on the subject of Japan's venture capital industry grew out of my interest in Japan's expanding financial role in the global marketplace. Although Japanese confidence has been undermined by political, economic and financial shocks, Japan's nascent venture industry has passed through its turbulent adolescence. The development of this particular subject matter was reflected in *Venture Japan,* a quarterly that tracks both venture trends and strategic alliances among Japan, North American and European companies.

There are many people responsible for this body of writing. I am grateful to Steven D. Bakalar, who generously introduced me to this important subject in his completed thesis. There are individuals like Carolynn Gandolfo and Koichi Itoh who provided unfailing encouragement and support for this research and development. Others like Dr. Yaichi Ayukawa, Tom Nishizawa, Stuart Laidlaw, Neeraj Bhargava, Art Spinner, Edwin Goodman, Shigeru Masuda, Burt Alimansky, Alan Patricof, Robert Brown, Arthur Mitchell, Marsha Monro, and Wenke Thoman provided gentle guidance and cooperation. I am also grateful for the generous contributions from Hiroaki Ueda, Hidemi Suzuki, Takeru Ohe, Shuji Honjo, Mark Oliva, and Ian Macmillan.

## viii  Acknowledgements

Although the venture capital industry is historically a closed community, particularly in Japan, where no one wants to reveal the size of their fund or investment portfolio, I managed to penetrate that silence and discover a few open doors and make new friends. My friend, Shoichi Fujikawa, the president of JAFCO America Ventures candidly shared information with me and discussed some of their strategies. Another Japanese who deserves a special 'arigato is Hideo Arakawa, the director of the Venture Enterprise Center in Tokyo. He was also most responsive to my urgent fax communications and telephone calls for additional research data. The completed database of Japanese venture firms is presented in a worksheet manner for practical usage.

I am also indebted to the time and effort spent on administration, data-gathering, fact checking and typing handled with speed and enthusiasm by Renzo Cella. Whatever deficiencies either in breadth or depth of subject matter, the sole responsibility for any shortcomings in the final product is all mine.

James W. Borton
New York

# Japanese Venture Capital: A Historical Overview

## James W. Borton

This book examines the growth of venture capital in Japan and the globalization of venture business.The distinguished authors of the following chapters reflect divergent points of view; however, they reveal valuable financing patterns, astute investment commentary and current venture trends in Japan. A number of diverse and dynamic events are taking place in the current world of venture business between Japan, the United States and Europe. The venture capital industry in Japan is continuing to emerge and is usually different from its U.S.and European counterparts. These differences and relationships are introduced in this prologue. The past scandals in the Tokyo Securities industry, coupled with a decline in profits at Japan's

x   Introduction

largest banks, has put some pressure on the capital adequacy of the Japanese financial institutions. Not surprisingly, even the pool of new U.S. venture capital declined to less than $2 billion in 1990, from the record high of $4.2 billion in 1987.

Nevertheless, in view of the deregulation of Japan's financial market and the increasing Japanese presence in global finance, the growth of venture capital in Japan seems a natural development. Since the early 1980s Japan's venture capital has progressed to its current investment of 1152.2 billion yen. Although this sum is still dwarfed by the U.S. total venture capital investment, its significant growth and its differences from U.S. venture capital are worth exploring. After years of aversion to risk-capital enterprises, Japanese banks and securities houses are becoming more involved in the venture capital business.

According to Japan Associated Finance Co. (JAFCO), the largest venture firm in the country, currently there are 104 venture capital firms operating in Japan. Increasingly, venture capitalists are diversifying into new sectors, such as services and retail. The integration of the U.S., European and Japanese markets offer cross-border investment opportunities for both the Japanese, American and European venture capitalist.

Venture capital is fundamentally different from traditional "passive" investment mechanisms that involve only financial resources; it entails a partnership between the venture capitalist (VC) and the entrepreneur. The VC provides funds and value added services such as operating and strategic advice at a time of risk and uncertainty when the entrepreneur needs both. In return for his or her investment of financial and human resources, the VC receives equity ownership in the company on terms that enable him or her to reap high returns through capital gains when the investment becomes liquid through a public offering of the stock or a sale of the company.

As defined in the United States, where venture capital was invented in the 1950s, venture capital investors receive equity ownership in young companies, with the objective of eventually achieving large capital gains. However, equity investments account for only 20 percent of the 783 billion yen that the Ministry of International Trade and Industry's Venture Enterprise Center (Kenkyuu Kaihatsugata

Kigyoo Shinkoo Shitsu) defines as the outstanding balance of cumulative venture capital investment in Japan. Various forms of debt that would not be considered venture capital in the U.S.—convertible bonds, warrant bonds, debenture bonds, and loans—account for the other 80%. The term "young" is arguably a relative term. In the U.S., a firm receiving venture capital, commonly known as a "start-up," is rarely more than 3–4 years old. However, Japanese firms may be up to 15 or more years old when they receive venture capital.

In Japan, the term *benchaa bijinesu* (venture business) was coined in 1972 by Japanese scholars and MITI officials to describe:

> ...small- and medium-sized independent companies with an active entrepreneurial spirit and aspirations for dynamic business expansion or the development of unique technological expertise and/or managerial know-how.

To summarize, venture business refers to knowledge-intensive and innovative small business.

The first Japanese venture capital boom occurred around 1972 and was represented by eight firms. These included Kyoto Enterprise Development (KED), the first Japanese venture capital firm, followed by Nippon Enterprise Development (NED), which was established by the Long-Term Credit Bank of Japan, the Dai-ichi Kangyo Bank and C.Itoh. NED was in turn followed by six other firms, including the Japan Associated Finance Company (JAFCO), an affiliate of Nomura Securities. Together these eight firms represented the first Japanese venture capital industry. Following a high growth period in the late 1960s, financial institutions had ample cash reserves and were looking for new investment opportunities. This experiment at venture financing was short-lived; however, as many portfolio companies failed to even get off the ground and most of the VC firms fell deep into the red. While some of the blame can be attributed to the dual oil shocks of 1973 and 1979 that created a dismal environment for small high-tech operations, the main problem was that the pioneer venture capitalists were high on spirit and low on experience. They neither sought advice from their U.S. counterparts nor invested directly in the U.S. to gain expertise. Also, these affiliates of large financial institutions lacked the autonomy to make the swift and decisive decisions needed in this time of economic crisis. Additionally, the

Ministry of Finance (MOF) imposed stricter regulations on industry, and accounting requirements for Tokyo over-the-counter (OTC) market registration were stiffened substantially. The original VC firms pared down their venture investments and focused instead on safer investments such as leasing, factoring, and consumer loans.

The second venture capital boom occurred around 1982. By this point, numerous changes had occurred in the industry. Financial liberalization in 1980 enabled Japanese venture capitalists to more easily receive capital and advice from the U.S. VCs, as well as to increase their own investment activity and experience level in the U.S. Investment partnerships were established for the first time, which provided for risk diversification. In addition, MOF provided a stimulus by relaxing the Tokyo Stock Exchange (TSE) Second Section and OTC listing requirements, and by liberalizing the use of warrants. Finally, VCs developed a rating system for venture businesses similar to the bank rating system for large corporations, which acted as an important signalling device to banks to facilitate loans and to potential customers to support sales.

Despite these favorable trends in the Japanese venture capital industry, this boom period bottomed out shortly after three companies "decorated with the label of venture business star" collapsed in 1986, creating a shock wave throughout the venture community. Many VCs shifted attention to foreign investment activity, taking advantage of the new era of the *endaka*, strong yen ushered in by the famous G-5 meeting in late 1985. But even in the absence of exchange rate revision, one could see that these venture businesses were precariously perched on the crest of an explosive wave. Kazuhiko Nishi, of Ascii Corporation, himself a well-known venture businessman, admitted that although the businesses with venture capital backing had the financial resources to support their fast growth rate, they suffered from the lack of management resources.

The current status of venture capital in Japan is more promising. As of July 1990, there were 100 venture capital firms in Japan. 32 are joint ventures of securities firms and regional banks, 22 are affiliated with securities firms, and four are classified as foreign. Of the firms with financial affiliation, the majority have equity ownership among regional banks, mutual loan and savings banks, cooperative associations and securities firms. This represents a major change from the past when the predominant form of organization was affiliation with only one securities firm.

There are three different ways to define venture capital funds: equity only, equity plus equity-like, equity plus equity-like plus debt. MITI's VEC uses the latter definition, since they have an obvious incentive to maximize the size of industry under their jurisdiction. According to Shoichi Fujikawa, president of Japan Associated Finance Company, JAFCO, JAFCO American Ventures, Inc., the vast majority of funds come from corporations. Manufacturers and financial institutions supply 90%, pension funds 5%, and individuals 5% of Japan's venture capital funds. By comparison, in 1986 pension funds supplied 50%, individuals 12%, and corporations 11% of U.S. venture capital funds.

Similar to the U.S., there is a fairly wide distribution of investment across industries. While in both countries the sector receiving the most venture capital funds is the electronics sector, a much greater percentage of total funds go into electronics in the U.S. than in Japan. Defined rather liberally by VEC as "Denki Kikai Kigu Seizoogyo," the electronics sector would include computer hardware and systems, telecommunications, and data processing equipment.

Japanese venture businesses can go public by registering their shares on the Tokyo Stock Exchange (TSE), First or Second Section, regional exchanges such as Osaka and Hiroshima, or the Tokyo OTC. Those that register on the OTC market sometimes jump up to a major exchange after a few years. In 1990, 49 firms had Initial Public Offerings (IPOs) on the major exchanges, and 86 firms went public on the OTC market. The gradual increase in OTC IPOs in the 1980s suggested that the OTC market was opening up and represents an increasingly viable cash out vehicle. It is interesting to note that over 50% of Japan's IPOs to date have involved service-sector ventures.

The real problem for both VC and entrepreneurs is the long wait until IPO time. 92% of the companies listed on the Japanese OTC exchange were 15 or more years old when they went public. Companies younger than 10 years comprise 42% of the U.S. NASDAQ market, compared to less than 1% in Japan. Thus, VCs who invest in early stage Japanese ventures have to wait a lifetime to cash out on

## Table 1
## Japanese Venture Capital Investment

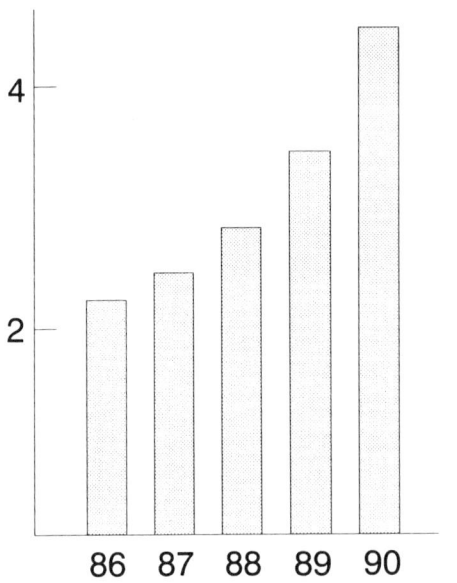

In ¥ 100 billions, as of end of March 1990

The balance of venture capital investment is expanding posting a 32% year-on-year gain to ¥ 474.6 billion as of the end of March 1990. Japan's largest venture capital firm is Japan Associated Finance Co., an affiliate of Nomura Securities Co., currently holding ¥ 131.6 billion in assets.

their investment, while the businesses themselves have to prove long-term staying power to reach the IPO level.

In general, the government can influence venture capital through tax policy, as well as grant and loan guarantee programs targeted at investors and/or entrepreneurs. Japanese tax policy is not particularly conducive to venture capital development. Corporations, which supply the bulk of Japan's venture funds, are saddled with a 50% capital gains tax that serves as a strong disincentive to invest with capital gains objectives. Both the corporate and individual income tax rate are steep as well. Japan lacks a corporate structure similar to the U.S.'s Subchapter S, which gives private companies with closely held stock partnership tax status and provides a partial shield from double taxation.

MITI's main supportive act has been the creation of Venture Enterprise Center (VEC), a non-profit organization, in 1975. VEC promotes the activities of venture businesses and venture capital companies in Japan. Along with collecting information on and providing analysis for the industry, VEC provides loan guarantees for small- and medium-sized firms for an "original project." VEC will guarantee up to $450,000 and 80% of the total project loan. As Japan is perhaps the ultimate collateral society, qualifying firms are relieved of the heavy collateral burden lenders impose. Hideo Arakawa, the president and VEC's advisory committee select about 30 out of the approximately 100 applications filed every year. While VEC is not charted to distribute loans or grants, its guarantee acts as a signalling device to lending institutions, reducing interest rates and increasing supply availability. (See Table 1.)

Yamaichi Finance Co. Ltd. has its own fund and is an excellent example of a Japanese equity fund. Yamaichi Uni Ven was created in 1982 by Yamaichi Securities with an initial capitalization of 200 million yen. Since establishment, it has set-up five partnerships locally and two overseas for equity-related investments. In general, the lifespan of a YUV partnership is 10 years. The company operates much like a U.S. partnership, raising funds from limited partners, collecting a 3% management fee, and taking 20% of the realized capital gains. As of December 1, 1990, Yamaichi Uni Ven Co. Ltd. merged with Yamaichi General Finance and Yamaichi Card Services to form Yamaichi Finance Corporation. Its North American operations, Yamaichi Finance America, Inc., is located in Los Angeles.

YUV has raised five funds totalling 28 billion yen. Their portfolio consists of 300 companies and 120 IPOs. Over 75% of their portfolio securities are either common or preferred stock. Roughly 90% of these investments are in Japan.

Ninety-six percent of YUV's investment funds comes from corporations and financial institutions. The split between Japanese and foreign firms in this category is about 50/50. According to YUV, Japanese pension funds consider venture capital investments as overly risky. Three percent of the funds comes from pension funds, all foreign. The remaining 1% comes from wealthy Japanese individuals.

YUVs investments run across a wide spectrum of industries. Between them, electronic parts/equipment and computer related investments account for 32% of the portfolio, followed in importance by information processing and semiconductors. YUVs portfolio companies have had a total of 37 IPOs, 32 in Japan and 5 overseas. Its rate of return on investment (ROI) is a closely guarded secret. However, an analysis of five IPOs from 1985 and 1986 reveals a high ROI, particularly in light of the fact that these investments were late stage and low risk.

YUV management does not hold any board seat in its portfolio companies, but through the Yamaichi network, it is able to add value by providing assistance in financial planning, marketing, initial public offering, technology transfer, personnel recruitment and the fostering of local and overseas strategic partnering.

YUV is typical of Japanese venture capital partnerships in several ways. First, historically it has been conservative in early stage investments. Its brochure claims that YUV's principal objective is supporting business "in their early periods of growth" but then goes on to state that it will invest in companies "expected to go public in the foreseeable future (3-7 years)." Given the long wait required before a seed/early stage venture can go public, however, what this really means is low risk, late round "bridge" financing that provides a platform for the IPO. In light of the Japanese market's astronomical P/E ratio, it appears that YUV can make a healthy internal rate of return on sound late round investments without having to bear much risk. Nevertheless, Yamaichi Finance is now making early stage investments.

Second, YUV is not driven solely by financial return; it has multiple objectives, of which ROI is one. One objective is technology transfer. Much of its investment capital comes from operating companies that are interested in gaining access to new technology. This could be a technology related to their core business or a technology facilitating diversification into a new area. One of the reasons why YUV established a Los Angeles branch office in 1984 was to assist in the transfer of technology from U.S. university research facilities to Japanese industrial concerns. Banks represent another major source of funds. They are interested in forming new banking relationships with portfolio companies, particularly ones that have gone public and developed a strong equity base. Another objective is underwriting fees. Since Yamaichi Securities underwrites YUV's IPOs, fees for securities services are considered together with capital gains in calculating potential profit from an investment.

## Analysis of Venture Capital Development

There are numerous differences between Japanese and the U.S. venture capital. Most Japanese venture capital firms are affiliates of banks and securities corporations, while over half of the U.S. venture capital firms are independent. Partnership funds only account for around 30% of equity-related venture capital investments in Japan, while about 80% of venture capital investments in the U.S. are accounted for by limited partnerships. The U.S. VCs and investors are interested primarily in financial return on investment. In contrast, most Japanese VCs and investors have multiple interests. This is a function of where the funds are coming from and who is administering them. In the U.S., over 70% of venture funds are supplied by pension funds, university endowments and individuals who are interested in one thing: ROI. 75% of U.S. funds are managed by independent partnerships whose income is based heavily on capital gains. In Japan, both the suppliers and administrators of capital have multiple objectives that makes VC development all the more difficult.

Out of these different motives, in part, comes two distinct ways of evaluating businesses. U.S. VCs see management quality as the most important factor. If you ask them to rank investment criteria, most will tell you "people are first, market size/technology is second, idea/technology is third." Even if the U.S. VC likes the product and thinks there is a good market for it, he knows it is the entrepreneur who must make things happen. In Japan, product technology is the most important factor. Technology has an immediate and tangible value, but the problem is that technology in itself is not enough to grow and sustain a business, which is the purpose of venture capital.

As for the organization of the venture capital firms, Japanese firms employ a corporate structure, most adopting vertical structure based on comprehensive team work. The common U.S. organization is horizontal and small-scale organization, based on the partnership system. The backgrounds of almost all Japanese venture capitalists are finance related, such as securities firms and banks.

Perhaps the most striking contrast between U.S. and Japanese venture capital is the relationship between the VCs and their portfolio companies. U.S. partnerships are active toward and maintain a close relationship with their companies. They provide more than just capital by adding value in ways other capital sources cannot. VCs help prepare business plans, advice on operations, personnel recruiting and strategic planning. Entrepreneurs trust the VC, rely on his or her expertise, and freely release company data to him or her. The flow of information is truly two-way.

In Japan, the relationship between venture capital and venture business is, in general, distant and hands-off. Japanese venture capitalists seldom sit on the board of their portfolio company. Part of this can be attributed to a fiercely competitive national consciousness, what the Japanese call *kyoosoo ishiki*. There is a fear that disclosed information will somehow find its way into the hands of competitors. In addition, Japanese VCs as a rule have a low reputation among venture businessmen. Entrepreneurs fear that VCs seek short-term profit at long-term expense, are greedy and will take over the company. The problem is that the vast majority of Japanese VCs do not have operating experience and therefore lack credibility with the venture business. On the other side, Japanese VCs have trouble getting information on potential portfolio companies from third party sources. Many do not trust the information they receive from venture businesses themselves. In general, networking with other venture capital firms in Japan is slowly progressing but compared with the United States this field needs more development.

A final difference between Japanese and U.S. VCs is the stage at which the VC invests into the target company. In the U.S., the value-added period constitutes the theoretical foundation for the concept of venture capital. It is characterized by high risk, financial peril and constant trouble-shooting. Although these so-called "seed/early stage" investments account for only 35% of all U.S. venture capital disbursements, they represent the heart of the U.S. venture capital system. This is the time when the VC can contribute significant value to the fledgling and often struggling company. In return for the value-added the VC gets cheap equity with the prospect of high returns if the company makes it to the IPO stage.

On the other hand, Japanese VCs put less than 5% of their total disbursements into seed and early stage investments. They avoid investing in the value-added period because they cannot add the value the company needs, and cannot justify taking the high risks associated with these investments. The risks are higher than in the U.S., in large part due to illiquidity. IPOs are more difficult to achieve, and acquisition still faces strong cultural resistance. By concentrating on the post venture/pre-public period, Japanese VCs are simply providing bridge financing to sustain the company unity it can issue an IPO.

One of the biggest challenges facing Japanese venture capitalists is how to develop a value-added concept appropriate for Japan. Added values offered by Japanese venture capitalists are primarily related to financial services, but this is changing as VCs are beginning to provide a broader spectrum of added values to their investees. Techno-Venture, a firm affiliated with foreign VC operations, offers new forms of value added and may represent the new wave of Japanese VC. Founded by Dr. Yaichi Ayukawa, Techno-Venture raised its third fund in 1986 totalling 3.5 billion yen. Through an alliance with Advent International, an affiliate of TA Associates (the world's largest venture capital firm) which serves as the coordinating mechanism for a network of funds around the world, Techno-Venture can offer Japanese venture businesses a number of unique services. First, it can call on the participation of various experts in diverse technical fields to evaluate businesses and to provide consulting services. Second, it can help find foreign export markets for Japanese products. Third, it can help Japanese venture businesses set up manufacturing facilities and/or subsidiaries in foreign markets and help them become "insiders." Fourth, it can introduce Japanese venture business to potential corporate partners for marketing/distribution and/or manufacturing alliances. Fifth, it can offer Advent's Hong Kong manufacturing facilities as a low-cost production site. According to Dr. Ayukawa, the Techno-Venture philosophy can be described by an international division of labor system called AMI (Advent Manufacturing/Marketing International), where research is performed in the U.S. development in Japan. Of course, these value-added services are not limited to Japanese venture business. Techno-Venture's non-Japanese portfolio companies also benefit from these services. From Techno-Venture's perspective, by providing potential acquirers for portfolio companies as an alternative to the IPO process, the international network increases the liquidity and reduces the risk of its investments.

Given this brief overview of the Japanese venture capital, what is the future of venture capital in Japan and what opportunities lie ahead? The business climate for the venture capitalist is favorable, considering the availability of new capital in Japan. Small firms that venture capitalists invest in have a strong heritage in Japan and have found a secure niche among Japan's corporate giants. The OTC market remains more attractive to small- and medium-size businesses, and the government's continued deregulation program with relaxed capital gains taxes and financial regulations offer other incentives to venture capitalists.

The following are the primary current trends involving Japanese venture capital firms.

- Japanese venture capital firms have set out to invest in U.S. and European high-tech small businesses.

- Cooperation between U.S., European and Japan venture capital firms is focusing on two aspects: fundraising and co-investing.

- Japanese venture capital firms are investing in Japanese subsidiaries of U.S. and European corporations, with the intent of taking them public in Japan.

- U.S. and European corporations are stepping up their investment in Japanese venture capital.

As for potential ventures in Japan, venture capitalists are diversifying into new sectors. These range from the service sector, such as software and retailing, to high technology and the leisure/recreation industry. Japanese venture capital firms are making efforts to provide the kinds of hands-on services created by the U.S. and European venture capital industry, despite pursuing a different model of venture capital than the typical continental venture capitalist.

While the U.S. and European venture industry has indeed experienced a serious downturn this past year, Japan's insurance companies are increasingly investing in start-up companies. Nippon Life, one of the largest in the world now has about 1,000 venture capital investments worth over 75 billion yen. Other large Japanese insurance companies like Dai-ichi Mutual Life now provide venture capital through a separate subsidiary. While there is a corporate contraction globally for investments in start-ups, Japan's corporate elite has shifted from investments in large-capital stocks to opportunities in the growth potential of small firms. While the majority of investment capital is now concentrated on later stage investments and management buyouts, Japan continues to distinguish itself in the streamlined venture industry.

# Chapter One

# The Internationalization of Venture Capital

## Alan Patricof

Until a few years ago, venture capital was as American as apple pie. We had the money, the entrepreneurial tradition, the over-the-counter market, and the tax structure to make venture capital practically an American monopoly.

Today, venture capital is increasingly a conduit for the flow of capital and ideas on a global basis. In 1989, according to the European Venture Capital Association, more money was raised for venture capital in Europe than in the United States—$5.3 billion versus $2.4 billion. Though activity is greatest in France and the United Kingdom, there are now venture capital associations throughout Western Europe. Japan has venture capital firms. And even China has started to think of ways risk capital can help solve its development problems.

## No Tradition of Risk

The wisdom of investing in European growth companies may seem obvious today, but less than ten years ago there was a mood of almost universal skepticism. There were a number of hurdles that had to be cleared before U.K. venture capital could flourish. Unlike the United States, the United Kingdom lacked a recent tradition of entrepreneurial risk taking. The idea of moving from job to job—not to mention leaving to start a new company—was frowned upon. And so it was throughout Western Europe. For decades, European industry had been dominated by the state, as in France and the United Kingdom, or by the banks, as in Germany. Size was rewarded over innovation. Lifetime employment was valued over entrepreneurship.

The Thatcher administration broke the mold. Margaret Thatcher enacted a number of tax concession measures to encourage entrepreneurial activity. At the same time, her administration developed the Business Expansion Scheme, which offered tax breaks to private citizens making investments in young companies.

Even more important was the creation of the Unlisted Securities Market (USM) in 1981. As "little brother" to the venerable London Stock Exchange, the USM nurtured the kind of young companies that venture capitalists thrive on. Then came the Nightingale Market, an OTC-style market with even more relaxed requirements. Suddenly, British entrepreneurs could tap the capital markets, and venture capitalists could recycle their investments.

Public markets are the key to why venture capital thrives in one country but not in its neighbor. Without a strong OTC-style market, venture capital cannot take root—even in the most highly developed economies. For example, French venture capital took off only after 1983 with the creation of the Seconde Marche.

If markets are an integral element, they are by no means the only one. British venture capital has also been spurred by legislation permitting stock option plans. Lower tax rates have increased the incentive to invest in young companies. Flexible labor regulations have made it easier to correct redundancies.

Other countries seem to be struggling to find the right mix. Germany, for example, has an OTC-style market, but companies that want to go public need the backing of banks, which have stringent requirements that discourage fledgling companies.

Japan, for all its status as a premier economic power, has only started its venture capital industry, and it remains to be seen if the regulatory and cultural problems that hamper its growth can be overcome. The OTC market in Japan, while liberalized in 1984, is still considered a less preferred option by Japanese investors, which makes taking a company public difficult. Start-up companies, meanwhile, have difficulty finding top-level management and labor because of the country's lifetime employment system. Even so, there are optimistic signs. In 1989, there were 73 initial public offerings in the OTC market, almost four times the number seen just two years ago and the market is expected to continue this growth.

## Venturing into 1992

Still, for institutional investors, non-U.S. venture capital opportunities overseas are likely to be found in Europe. That is particularly true given the European Community's plans to remove financial and market barriers by December 31, 1992. With a population of over 320 million, the European Community has nearly as many people as the United States and Japan combined, and its GDP is $4.9 trillion, versus $5.1 trillion for the United States and $2.8 trillion for Japan. The possibilities for venture capitalists are intriguing. Europe is approaching the point at which a product can be financed by one country, manufactured in another, and marketed in a third. Companies will not be merely French or Italian, any more than companies in Philadelphia or Los Angeles are merely Pennsylvania or California companies. One recent example of this trend is European Silicon Structures, a company that is incorporated in Luxembourg, headquartered in Munich, with research facilities in England and a factory in South Korea.

But for venture capital to extend beyond the United States and Europe, several things must happen:

- Countries must have a strong OTC market.
- Governments must liberalize their tax structure and deregulate industries.
- Labor rules must be more flexible.
- The concept of the entrepreneur must become part of the societal framework.

It is not going to happen tomorrow, but we should not totally discount the possibility. After all, look at what has happened to Europe in a few short years. In the United Kingdom there are as many as 100 initial public offerings (IPOs) a year, up from a handful ten years ago. France now has an average of 40 IPOs a year, compared to two or three ten years ago. As other countries continue to realize the need for entrepreneurial innovation, they will look toward the example set in Britain. (See Table 1.1.)

## The Japanese Look Abroad

While Japanese investors have been slow to back venture capital at home, they have long participated in U.S. venture funds, and now they are turning to Europe as well. A recent example of this was the investment by Sumitomo Corporation and Nippon Investment Finance in Euroventures BV, a venture capital conduit formed by seven major European companies.

The obvious reason for the interest in venture capital is return on investment. In the United States one can realistically hope for a sustained 20% annualized rate of return from venture capital. A good rule of thumb is: Look for 300 to 600 basis points more return on venture investments than on ordinary equity investment. Over ten years, that premium can result in a significant increase in capital.

## Table 1.1
### Venture-Backed IPOs: 1982-1989

|      | London Stock Exchange | Unlisted Securities Market | Nightingale |
|------|----------------------|---------------------------|-------------|
| 1982 | 1                    | 7*                        |             |
| 1983 | 2                    | 20                        |             |
| 1984 | 8                    | 14                        |             |
| 1985 | 7                    | 12                        |             |
| 1986 | 19                   | 22                        |             |
| 1987 | 22                   | 10                        | 4*          |
| 1988 | 18                   | 16                        |             |
| 1989 | 7                    | 11                        |             |

*First year for this market.

While we have not yet gone through a full ten-year cycle in Europe, I would expect returns to be comparable.

Institutional investors have turned to international venture capital to diversify their portfolios—not only in terms of assets, but also in terms of economic exposure by domestic economies and currencies.

Equally important, if less obvious, is the stock market crash. The traditional objective of most institutional investors was the greatest return on the most liquid investments. But the lesson of October 19, 1987, is that liquidity can be illusory. In a matter of a few days, public equity values in the world's stock markets fell by a quarter to a third. Venture portfolios were far less severely affected because the psychological whims of the public market are neutralized in a non-marketable portfolio.

Japanese participation in venture capital, while it goes back to the early 1980s, really took off in 1985 when Mr. Tetsuo Imura raised $28 million for the Orien Venture Capital Fund. This fund, which was formed as a joint venture between Vista Ventures, a U.S. venture capital firm, and Mitsui Trading Company, included as participants 21 Japanese corporations, including the following list of companies:

| | |
|---|---|
| Bank of Tokyo | Nippon Life |
| Chugai Ro | Nippon Steel |
| Fuji Bank | Nitta |
| Fuji Photo | Onoda Cement |
| Idec-Izumi | PNB (Malaysian National Bank) |
| Japan Tobacco | Sapporo Breweries |
| Kanto Natural Gas | Shimizu Construction |
| Mitsui Taiyo Kobe Bank | Takada |
| Mitsui Toatsu Chemical | Tokyo Seimitsu |
| Toppan Printing | Toyo Engineering |

In 1988, Mr. Imura led the formation of Orien II for $50 million in capital that included 26 Japanese corporations, 19 of which came from the original group. In addition, there were at least half a dozen participants in both funds from American and European corporations.

Recently, Hambro International launched the fundraising for Hambro International Equity Partners III (HIP III), and it was initially capitalized with $20 million from Hambro Bank and Mitsubishi Corporation who will be assisting in the fundraising. When fully capitalized, the fund is expected to hit $150–$200 million. HIP III is intended to take 20%, or greater, stakes in U.S. biotech, software and other high-tech companies. The capital for this fund is intended to come from Japanese corporations who will be given the opportunity to form strategic partnerships with the portfolio companies if they are participants in the fund and also to invest directly in the portfolio companies.

Mitsubishi's role also will include the development of strategic alliances, the evaluation of markets and the pursuit of liquidity on a global basis. According to Hambro, multinational partners such as Mitsubishi, which have access to both capital and international markets, provide the resources to help start-ups compete globally.

Nomura Securities International has been actively assisting U.S. venture firms find Japanese investment partners. Most recently, Nomura acted as the investment banker for the International Bioscience Fund-Japan, a venture partnership established by Oxford Partners and Matuschka Venture Partners that will invest in emerging growth companies in the biosciences alongside the International Bioscience Fund. The International Bioscience Fund has received commitments from U.S. and European investors and expects to close with $50 million. The International Bioscience Fund-Japan has $20 million committed from Japanese investors.

Nomura Securities was also very effective in securing Japanese investors for Dominion Ventures' newest fund, Dominion Fund II, which closed in April 1990 with $42.5 million. About 25% of the total capital is Japanese funds. Japanese investors include Mitsui USA, Daiwa Bank and Japan Tobacco. The Fund will focus on early stage investments in the high-technology area.

Continental Bank through its Japanese affiliate, Continental Capital Markets, also has had some success in raising venture funds for U.S. firms. According to Mr. Qunio Takashima of Continental Capital Markets, his firm recently helped raise venture funds totalling about $15-$20 million for several firms including Summit Partners, a Boston-based firm and Golda Thoma Cressey of Chicago. The funds came from Far East sources, primarily from trading companies.

While several of the Japanese organizations that have become involved in U.S. venture funds are financially oriented, particularly the banks and insurance companies, many of the firms, such as Mitsubishi, have particular related interests and hope to obtain a window on technology in order to expand their commercial activities. Through the use of investments in the venture business, these companies were eager to become aware before everyone else of emerging technologies and through that exposure either to create investments in the specific portfolio companies or to initiate joint ventures or distribution agreements.

Not surprisingly, the largest percentage and dollar amounts of direct investments have been in the computer hardware and software field, primarily in the California area. There is, however, an emerging interest in biotechnology as evidenced by Oxford Partners and

## Table 1.2
### Japanese Investment in U.S. Venture Businesses

| | Minority Equity Investments* in U.S. Firms | | | Investment in U.S. Venture Funds | |
|---|---|---|---|---|---|
| Year | Number of Investments | Dollars Invested (Millions) | Average Investment Size | Number of Investments | Dollars Invested (Millions) |
| 1983 | 11 | 7 | .6 | 4 | 18 |
| 1984 | 15 | 44 | 2.9 | 14 | 28 |
| 1985 | 15 | 42 | 2.8 | 24 | 31 |
| 1986 | 20 | 142 | 7.1 | 9 | 33 |
| 1987 | 49 | 151 | 3.1 | 8 | 14 |
| 1988 | 47 | 176 | 3.7 | 22 | 46 |
| 1989 | 60 | 320** | 5.3** | 16 | 54 |

*Publicly reported, by industrial corporations only.

**Includes a $100 million investment by Canon Inc. in NeXT Inc., excluding the $100 million investment, average investment size for 1989 is $3.7 million.

Note: Japanese investment statistics are periodically updated to reflect additional information made publicly available by corporations.

Source: Venture Economics Inc.

---

Matuschka Partners' new International Bioscience fund as well as a fund Daiwa Securities is raising, the Gilliam/Daiwa Life Science Fund, focused on aging disorders in the life science field. Japanese investors are expected to commit about 50% of the capital to this fund, with the other half coming from U.S. institutional investors.

It appears, based on the growing participation of Japanese corporations in U.S. venture funds, that many industrial corporations first make their investment in venture funds until they develop some familiarity with the field and begin the process of engaging in corporate partnering or forming strategic alliances. After that, they begin to make more direct investments.

This approach is supported by the data collected by Venture Economics (see Table 1.2) which shows that Japanese corporate investment in U.S. venture funds has increased steadily since 1983, while the growth of Japanese investments in U.S. companies picked

## Table 1.3
## Japanese Minority Equity Investments in U.S. Small Firms in 1989 and 1990

| Japanese Company | Invested (%/MM) | Business Area | U.S. Small Firms |
|---|---|---|---|
| Kobuta Ltd. | $10 (15% stake) | CAE software | Rasna Corporation |
|  | $12 (25%/J.V.) | Optical Storage | Maxtor Corporation |
|  | $20 (NA) | Graphic Supercomputer | Ardent Computer |
|  | $9.9 (14%) | Biopesticides | Mycogen |
|  | Undisclosed Amount | Image compression processors | C-Cube Microsystems |
| Yokogawa Electric Corp. | $13.5 | Supercomputer | Supertek Computers |
| Sumitomo Chemical | $NA (20%) | Insecticides | McLaughlin G King Co. |
|  | $10 (NA) | Pharmaceutical | Regeneron Pharmaceuticals Inc. |
| Sumitomo Metal & Industry | $36 and Debt | Magnetic & Materials | Three firms: Biomagnetics LTX Mosaic Systems |
| Canon | $100 | Computers | NeXT |
|  | $4 | Graphic Software | MetroLight Studios, Inc. |
| Kobe Steel | $20 (7%) | Data Storage Disks | Komag |
| Toyo Engineering | $250,000 | Robot developer | Remote Technology |

Source: Venture Economics

up dramatically after 1986. These investments soared to $142 million in 1986 from $7 million in 1983. In 1989, Japanese investments in U.S. firms reached $320 million, including a $100 million investment by Canon Inc. in NeXT Inc. Overall, the Japanese make direct investments in U.S. firms to gain access to technology, to provide the ability to diversify into new business areas, and to form strategic business opportunities.

An interesting example of such diversification is Kubota Ltd., a Japanese agricultural equipment maker that has made direct investments in U.S. firms to help it expand into the computer business. (See Table 1.3.) In 1989, Kubota made five investments in U.S. companies including Rasna Corporation, a developer of computer-aided engineering software; Ardent Computer Corporation, a graphics supercomputer maker; and C-Cube Microsystems, a San Jose, California-based start-up that manufactures image compression processors, software and developers' platforms. With the increasingly global nature of the high-technology markets and the desire for multinational corporations to expand geographically rather than through business line diversification, small U.S. firms are finding it necessary to internationalize sooner and it appears that the easiest way to accomplish this is by working closely with a Japanese partner.

My colleagues and I, at Alan Patricof Associates, realized some time ago that Japan was a vital market to be tapped and one that would require careful nurturing. Just over three years ago, during fundraising for APA Excelsior III, we approached various Japanese institutions and were successful in securing $15 million from two major Japanese financial institutions. And now, we continue in our efforts to establish solid, long-term relationships with major players in the Japanese market. We are pursuing this market more aggressively as we look ahead to the 1990s and beyond and see the internationalization of the industry take hold.

All things considered, international venture capital will continue to expand. A number of other countries are beginning to create the atmosphere needed for venture capital to grow. The unification of Germany and the growth of German venture capital organizations certainly makes this country an exciting prospect for the 1990s. In 1989, Germany was one of more active markets for venture investing.

Alan Patricof Associates closed a Pan-European fund this year totalling just over $400 million that includes a fund in Germany for investments in German-speaking countries.

The U.K. and France continue to be very active venture capital markets. In 1989, the U.K. raised a number of very large funds specializing in buyout and other later-stage opportunities and the majority of the funds raised in France was for buyout, buy-in and turnaround situations. Italy had the third largest pool of funds raised in 1989 but continues to encounter barriers to the development of the venture capital industry such as legislative constraints. A bright spot in the Italian market is the amount of cross-border funds that have found their way into the country, primarily from U.K. groups.

Although each European country has carved out its own unique identity in the venture capital industry, we are seeing a rising trend toward internationalization fueled by events such as the fall of Eastern Europe and the opening of those markets, the unification of Germany, the approach of 1992, a rising number of U.S. investors in European funds, and an increase in Pan-European and cross-border funds.

Venture capital has always been an extremely flexible tool for raising and investing capital in the technologies, industries, and innovations that help economies grow. During the 1990s, we will have ample opportunity to apply those techniques nearly anywhere in the world, particularly when events throughout the world continue to make a global business outlook an essential commodity.

*Mr. Alan Patricof is Chairman of Alan Patricof Associates, Inc., an affiliate of THE MMG PATRICOF GROUP.*

# Chapter Two

# Japanese Venture Capital in the United States

### Shoichi Fujikawa

The focus of this article is a description of the Japanese venture capital industry, and the activities of Japanese venture capital firms in the U.S. Like the U.S., the venture capital industry in Japan is very small compared to other financial fields. The reason why people pay so much attention to venture capital is its uniqueness and challenge.

JAFCOs venture capital investments represent almost half the total capital committed in Japanese venture capital partnership funds. Also, JAFCO is one of the few Japanese venture capital companies that has offices in the U.S. and is actively involved in U.S. venture capital investments. Although this chapter bases its information mostly on JAFCO, I hope these observations represent the general Japanese venture capital industry, and its activities in the U.S.

JAFCO was established in 1973 by three prominent financial companies: Nippon Life Insurance, Sanwa Bank and Nomura Securities. In 1972 and 1973, the first so-called "venture capital boom" occurred in Japan. The success of the American venture capital industry became known in Japan, and several major Japanese financial companies established venture capital subsidiaries, based on the U.S. model. In terms of timing, JAFCO was the third venture capital start-up in Japan. From the beginning, Japanese venture capital companies have had a close relationship with major financial institutions such as security houses and large banks.

The first venture capital boom ended quickly, due to the recession caused by the first oil crisis. Most venture capital companies stopped their investment activities, or changed their businesses into leasing or factoring. During that difficult time, JAFCO continued its venture capital activities, although on a small and limited scale.

In 1982, JAFCO set-up Japan's first venture capital partnership fund, after studying the U.S. system. Before then, most Japanese venture capital companies made investments from their own equity capital and/or bank loans. As a result, their activities were limited.

Since 1982, JAFCO has set-up 36 partnership funds. The total amount now of money committed to these funds is almost $1.35 billion. JAFCO now has about 350 employees, in 36 branch offices and joint ventures throughout Japan. The joint ventures are usually with regional banks.

Most U.S. venture capital companies consist of general partners and associates. Usually the total staff is less than ten people. The reason why JAFCO and other Japanese venture capital companies are so large and formally structured is that most of their portfolio companies are more mature firms than the usual U.S. venture capital portfolio company.

For example, more than 90% of JAFCOs portfolio companies have been in business for 10 years or longer. They are usually profitable when JAFCO first invests. Actually, JAFCO is looking for a better term to describe its activities, since "venture capital" is not exactly correct.

Even though the Japanese venture capital industry has grown rapidly since 1982, its size and scope is still smaller than in the U.S. Please note that these statistics are rather old, and the current Japanese growth rate is now higher than that of the U.S., so the gap between Japan and America has become smaller than shown here. (See chart: Venture Capital Statistics 1989 U.S. vs. Japan.)

The U.S. venture capital industry is not only larger in size, but is more diversified than in Japan. Almost all Japanese venture capital investments are described as minority shareholding in mid-market companies. There are very few start-up investments, and almost no LBO or MBO activities by Japanese venture capital funds at the moment.

The attached chart shows the ranking of Japanese venture capital companies by the amount of capital managed. (See chart: Funds Managed by Japanese Venture Capital Firms {1989}.)

Although these statistics are dated, the actual company rankings have not changed much.

JAFCOs share was more than 40% in 1987, and we think it is now almost 50% of all capital invested in the Japanese venture capital industry.

Seven of the top ten venture capital firms are affiliated with, or subsidiaries of, major securities houses. Techno-Venture is the only firm that has the same kind of organization as an American venture capital firm. Most of their funds are invested in prominent U.S. venture capital funds.

Japan Asia Investment Company, Ltd., is a venture capital company specializing in investments in Asian countries, such as Thailand, Malaysia, Indonesia, Singapore and the Philippines. The Japanese government provided about half of their initial funds. The president, Mr. Imahara, was the former president and chairman of JAFCO.

Schroders PTV is actually the only active foreign venture capital firm in Japan. The general partners are Japanese but their money comes from Europe and America.

JAFCO has 170 public companies and 650 private portfolio companies as of March, 1991. Each year, JAFCO makes around 100 to 170 new or follow-up investments. The investment process follows a systematic plan. The first step is finding a potential investee. We have a staff of 110 employees who work on identifying candidates, both by ourselves, and through introductions from our joint venture regional banks. One important qualification to be a candidate is the potential to go public within two to five years.

The so-called "finding teams" negotiate with the companies and prepare financial plans.

The second step is the investigation, done by a separate "investigation team" that visits the candidates and prepares research reports on them. Finally the deal is approved by an investment committee, consisting of JAFCOs senior officers.

The criteria to be a public company on the Japanese stock market is quite different from that of the U.S. The most important criteria is that a company should have a net profit before tax of $2 million or more when it goes public. Therefore, emerging high-tech or biotech companies that are still losing money cannot go public on the Japanese stock market.

JAFCO America was established in 1984 at Sand Hill Road in Menlo Park, California. The first mission was to raise a fund from North American institutional investors to invest in Japanese small businesses. We raised $36 million in 1985 from 16 prominent investors.

JAFCO began to invest in American small businesses after 1985. In the first three years, we did venture capital investments on a trial and error basis. We had three failures in the first five investments. The remaining two are so-called "living dead" companies. Therefore in 1988, after opening our New York office, we developed a more focused strategy for our U.S. investments. This strategy is clearly defined in the charts entitled "Venture Capital Investments" and "JAFCOs Preference of Private Equity Investment in the U.S." The chart entitled JAFCOs Selected U.S. Investments" shows JAFCOs investment companies in the past two years.

The biggest lesson we have learned from the past experiences in the U.S. is that successful venture capital investments require the most talented and experienced people. Even many American venture capital companies have had difficult times in the past seven years, especially after Black Monday in 1987.

To get this expertise, we prefer co-investments with prominent U.S. venture capitalists, and investments in companies already backed by venture capitalists. Also, we focus on later-stage investments.

As you can easily imagine, JAFCOs investments in Japanese companies would be considered later stage or mezzanine investments by U.S. standards. This is why JAFCO has a very high success rate. More than 70% of our portfolio companies go public.

JAFCO has the same kind of investment policy in the U.S. That is the main reason why we focus on later-stage investments. As for choice of industry, we prefer technology-oriented companies. This is because they have more business connections with Japanese companies and we can understand them better than non-technical companies. JAFCOs advantage exists in our wide ranging connections in Japan. We can use our networks in Japan to increase the business opportunities for our U.S. portfolio companies. For example, we are now finding distributors in Japan for several of our portfolio companies.

We have three major sources of introductions for our deals:

- U.S. venture capital companies.

- Other U.S. sources such as financial institutions and law firms.

- Finally, Japanese business corporations, which are seeking technology or marketing rights from U.S. companies.

Most Japanese companies are willing to pay money for technology licensing or Japanese marketing rights but are very reluctant to make equity investments in these firms. So, Japanese companies introduce JAFCO as an equity investor in the U.S. companies they discover.

We are still in the investment stage and have not harvested our portfolio companies yet. So, we have had very few actual returns from our investments so far. It is still too early to judge our strategy.

But, I feel that our strategy is working out very well. A lot of U.S. companies welcome JAFCOs investment not only because of our money, but because of our strong networking capabilities in Japan.

In the future, I believe JAFCO can expand our scope of venture capital investments in the U.S. such as start-ups, LBOs and other activities.

I'd like to take this opportunity to explain the other business of JAFCO America. We call it the "IPO" business. We are actively involved in recommending to American companies that they consider taking their Japanese subsidiaries public on the Tokyo OTC market. JAFCO makes investments in these subsidiaries before they go public, and assists them in preparing for the IPO.

This is still a new concept. But, there are already 18 U.S. subsidiaries listed on the Japanese stock market. The most successful case is Levi Strauss. They currently enjoy a very high PE ratio of more than 70 times earnings, on the Tokyo OTC market. Their business in Japan is expanding rapidly, and part of the reason for their success was going public.

It is a unique, and sometimes strange, idea for U.S. companies to take their subsidiaries public. However, we are finding that more and more American companies are understanding the Japanese stock market, and are realizing the benefits of this strategy.

Going public in Japan means not only getting funds from the public, but also joining the elite ranks of the best known companies, which creates much credibility and recognition from other businesses and the general public.

Banks see venture capital financing as a strategically significant means of strengthening their retail business. The growing number of medium-sized Japanese companies that are interested in listing their shares on the stock exchange are attracting the attention of securities firms. In fact, Industrial Bank of Japan, one of the strongest in big corporate transactions, is anxious to enter investment banking because of the relationship between two recent trends: the increasing number of venture capital firms, and the boom of companies going public on the OTC market.

Figure 2.1 shows a number of public companies in Japan and the U.S. In Japan, the Tokyo stock exchange has almost the same number

## Figure 2.1
## The Number of Public Companies

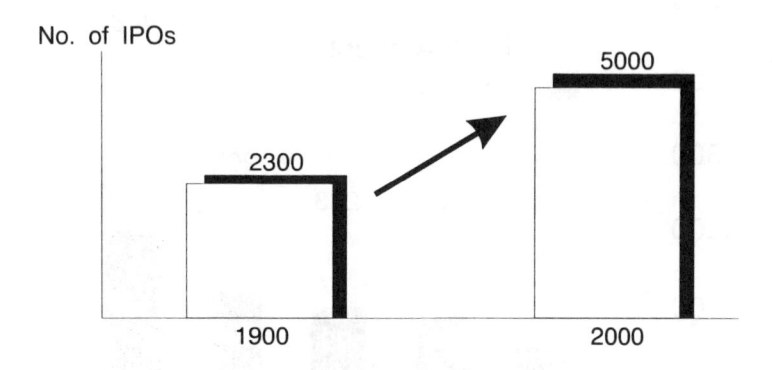

```
No. of IPOs
                                    5000
            2300
                         ↗

            1900                    2000
```

The number of public companies in Japan is expected to reach 5,000 by the year 2000. Many small- and medium-sized companies have a strong desire to go public. In addition, the role of Japan's over-the-counter market will become more important in Japanese capital markets.

In October 1991, the JASDAQ System was introduced. This automatic over-the-counter quoted system will help increase trading volume in this market.

---

of listed companies as the New York stock exchange, but a higher market capitalization. In contrast, the Japanese OTC market is still very small, although it is growing rapidly. In 1990, we had 86 new listings on the OTC, and we are expecting more than 100 new listings on the OTC this year. Investors have begun to view OTC companies as good investment targets.

JAFCO is now preaching the "IPO" business. The U.S. government is demanding that the Japanese government make a lot of changes to the Japanese economic structure. But, U.S. companies can

## Figure 2.2
### Total Amount of Investment Enterprise Partnerships

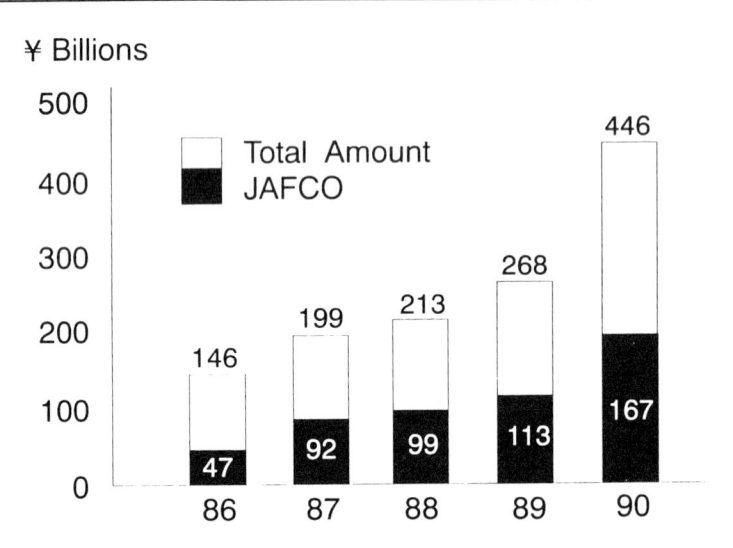

¥ Billions

| Year | JAFCO | Total Amount |
|---|---|---|
| 86 | 47 | 146 |
| 87 | 92 | 199 |
| 88 | 99 | 213 |
| 89 | 113 | 268 |
| 90 | 167 | 446 |

Investment Enterprise Partnerships (IEPs), introduced by JAFCO in 1982, have greatly expanded Japan's venture capital business. The total amount of JAFCO IEPs now exceeds 160 billion yen, demonstrating that JAFCO, Japan's largest venture capital firm, is continuing to grow.

---

take advantage of the Japanese economic structure, including low-cost capital, by taking their subsidiaries public in Japan. Some U.S. companies are doing this very quickly. I believe that this kind of mutual involvement in each other's economy will allow us to attain a true globalization of business and the development of a world economy.

*Mr. Shoichi Fujikawa is President of JAFCO, America Ventures.*

## Figure 2.3
## Number of IPOs by Year
## Japan and JAFCO Group

JAFCO was established with the backing of Nomura Securities and other leading Japanese financial institutions, including Nippon Life Insurance and Sanwa Bank 18 years ago, and the company remains Japan's leading venture capital company.

American venture capital companies are more centered around the individual. On the contrary, Japanese venture capital companies have large organizations with numbers of staff who have various expertise, and support the growth of many investee companies.

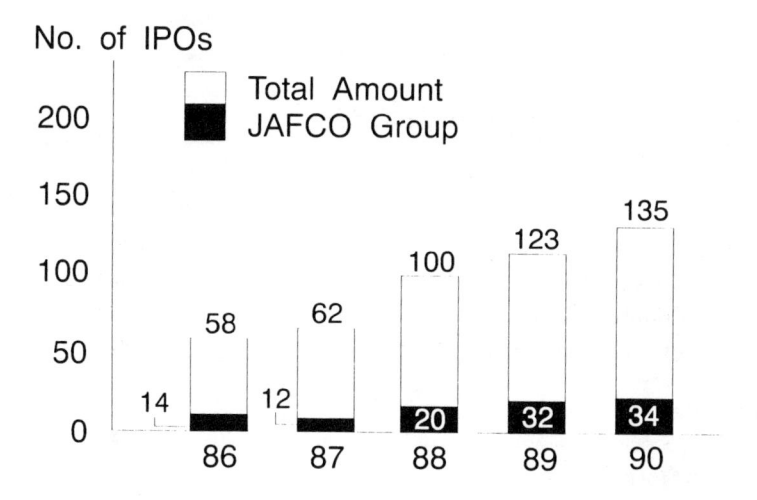

No. of IPOs

| Year | Total Amount | JAFCO Group |
|---|---|---|
| 86 | 14 | — |
| 87 | 58 | — |
| 88 | 62 | 12 |
| 89 | 100 | 20 |
| 90 | 123 | 32 |
|    | 135 | 34 |

More than eight hundred promising companies have been invested in by JAFCO. In 1989, there were 123 IPOs in Japan, and in 1990, there were 135.

By the end of 1990, JAFCO had helped a total of 158 companies go public.

## Figure 2.4
### Venture Capital Funds In Japan ($ Millions)

|  | 1985 | 1991 |
|---|---|---|
| JAFCO | $235 | $1,375 |
| Japan Asia Investment Corp. | N/M | $510 |
| Nippon Investment Finance | $85 | $390 |
| Yamaichi Finance | $80 | $200 |
| Nikko Capital | $45 | $190 |
| All Others | $205 | $185 |
| Total | $650 | $3,480 |

## Figure 2.5
### Number of Public Companies (1990)

| Japan | Number of Companies |
|---|---|
| Tokyo Stock Exchange | 1,627 |
| Regional Stock Exchanges | 444 |
| Over-the-Counter Market | 342 |
| Total Number of Public Companies | 2,413 |
| **United States** | |
| New York Stock Exchange | 1,769 |
| American Stock Exchange | 863 |
| NASDAQ | 4,131 |
| Pink Sheets | 17,000 |
| Total Number of Public Companies | 23,876 |

**Figure 2.6 IPOs in Japan**

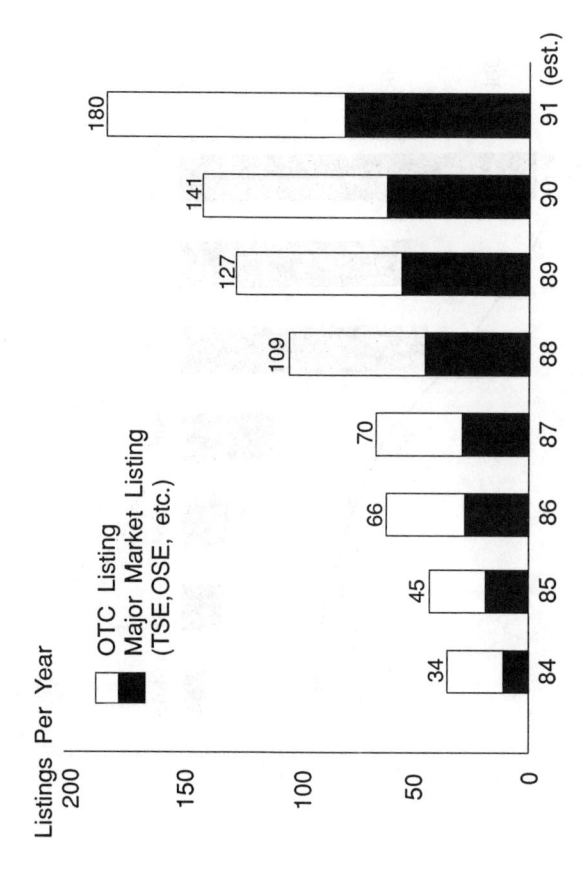

## Table 2.7 Number of OTC Listed Companies is Growing

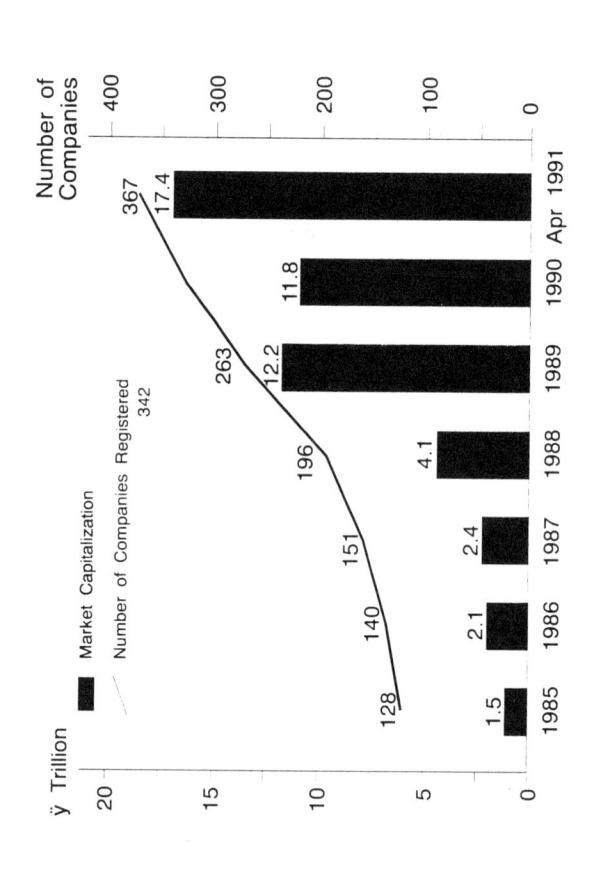

¥ Trillion

- Market Capitalization
- Number of Companies Registered

| Year | Market Capitalization (¥ Trillion) | Number of Companies Registered |
|---|---|---|
| 1985 | 1.5 | 128 |
| 1986 | 2.1 | 140 |
| 1987 | 2.4 | 151 |
| 1988 | 4.1 | 196 |
| 1989 | 12.2 | 263 |
| 1990 | 11.8 | 342 |
| Apr 1991 | 17.4 | 367 |

# Chapter Three

# The Euro-Japan Finance Network

## Giorgio Tellini

*The Nomura Group's prime institutional investor, Japan Associated Finance Company, JAFCO, is the new shareholder in SOFIPA, the Italian venture capital company, controlled by Mediocredito Centrale. SOFIPA's fully paid-up share capital has been raised to 120 billion lire.*

Societa Finanziaria di Participazione S.p.A. (SOFIPA) is a leading venture capital investment company in Italy, particularly active in development capital, and managing over IL 180 billion. At the end of April 1991, SOFIPA had more than IL 88 billion invested at cost in a total of 38 portfolio companies.

Centrale, the Italian public statutory credit institute that remains its principle shareholder, owning 51% of the share capital. Significant stakes are also held by Cassa di Risparmio di Torino, Mediocredito Lombardo, SAI S.p.A. and ACMER Sa (Banque Worms). Other shareholders include GIMV, the Belgian venture capital company, and a number of regional banks and private companies.

## Company Activity

SOFIPA acquires minority equity participations in medium-sized companies or subscribes to convertible bonds. While it is not company policy to take an active role on the day-to-day management of its portfolio companies, SOFIPA expects to be consulted on major issues affecting the companies' activities and thereby to play an effective role in the definition and implementation of their development strategies.

SOFIPA acts as a professional minority shareholder available to investee companies for consultancy on financial, administrative and organizational problems. In this way, SOFIPA gives added value to companies, over and above the equity participation.

Furthermore, mergers and acquisitions services are supplied both to the portfolio companies assisting them with expansion strategies, and to other Italian and foreign companies. As well as providing development capital to medium-sized businesses, SOFIPA also participates in management buyouts and buy-ins. These forms of financing, in fact, account for 59% of investment activity.

SOFIPA also takes part in equity capital restructuring (or secondary purchase transactions) by replacing minority shareholders that have no further interest in supporting the company's development. Capital restructuring accounts for an added 30% of SOFIPAs investment activity, while investments in new, or start-up, ventures account for 11%. SOFIPA also brings finance and other direct assistance to companies seeking a stock market listing and will underwrite issues.

In the past SOFIPA investment policy has been primarily focused on Italian companies from a wide range of industrial sectors. Since the formation of SOFIPA, over IL 122 billion have been invested in a total of 57 companies. Eighty-seven percent of the amount has been invested in Italy and 13% abroad.

Divestments to date stand at 15 for an amount of IL 49 billion at sales value. Two companies were floated: Gewiss S.p.A. in 1988 and Costa Crociere S.p.A. in 1989. Their market capitalization was about IL 220 billion and IL 191 billion respectively by May 11, 1991.

Among the most recent divestments by trade sale, one case is particularly notable, namely Gram S.p.A., which has been sold to Kellogg's corporation.

Through its Milan based subsidiary, SOFIPA Intermediazione S.p.A., SOFIPA also provides services and financial instruments to support corporate liquidity management and help control exchange rate, interest rate and portfolio risks.

There are a total of 14 investment executives at SOFIPA's Rome and Milano offices, under the leadership of Chairman Rodolfo Banfi, and General Manager Giorgio Medici.

The Japan Associated Finance Company Ltd. (JAFCO)—Japan's leading venture capital investor—joined the ranks of SOFIPAs shareholders when the share capital was raised to 120 billion lire, fully subscribed and paid-up.

JAFCO—incorporated in 1973—has an investment portfolio currently standing at 1,400 billion lire. One-hundred sixty-five investee companies have gone public, of which 13 listings are outside Japan. JAFCO operates throughout Japan, as well as in the United States (New York and San Francisco), Hong Kong and Singapore, and has one European branch in the U.K. (London). Twelve investment operations have been concluded in Europe to date, for a total amount of 23 billion lire.

The SOFIPA holding is JAFCOs first investment in Italy, and is a strategic step to take advantage of the possibilities for joint investments and working relationships in an Italian and European market with great development potential.

## THE NETWORK

SOFIPA has an international network of contacts and alliances that play a decisive role in identifying investment opportunities and co-investing in syndicated deals.

The network includes:

- three SOFIPA shareholders, ACMER (Banque Worms) in France, GIMV in Belgium and JAFCO in Japan;

- Hill Samuel Bank and ECI ventures in the U.K.;

- Siparex S.a. in France (Lyons);

- Siparex Participations S.a. in Switzerland, through a 6.5% stake;

- Euroventures BV in the Netherlands, in which SOFIPA holds a 1.2% stake;

- Societat Catalana de Capital a Risc, a Barcellona based venture capital company in which SOFIPA has a 2.4% interest.

SOFIPA expects to expand its network in the European countries and is particularly focusing on a relationship in Germany. New associations are also envisaged in the U.S., over and above the relationship with O.P.I. (Overseas Partners International), the management company of Columbus II, a closed-end fund based in Delaware for investments in Europe and the U.S. in which SOFIPA has also invested.

In South East Asia, SOFIPA invested in Asia Pacific Ventures, a closed-end fund fostered by Asian Development Bank and the Nippon Investment and Finance Co. for investments in the bank's member countries.

In Italy, SOFIPA has its banking and financial shareholders to provide an extensive network of relationships and information. As well as providing a valuable source of deal flow, this network enables SOFIPA to help its Italian portfolio companies to find international partners and also to assist foreign companies wishing to invest in Italy.

*Mr. Giorgio Tellini is Chief Executive Officer of SOFIPA.*

# Chapter Four

# The Promotion of Venture Business and the Venture Capital Industry

## Hideo Arakawa

Venture business in Japan dates from the 1960s. Although venture business is not yet clearly defined, in a word, venture business refers to knowledge intensive and innovative small businesses.

In the latter half of the 1960s, unique venture businesses were created, and received much attention from the journalists as a "venture boom."

The term "Venture Business" was coined in Japan. Professor H. Nakamura and T. Kiyonari mapped out the actual concept of the term in 1970. In a word, it refers to knowledge-intensive and innovative small business. In the interim report of 1984 on "the problems of venture businesses" by the Venture Business Study Group of Small and Medium Enterprise Agency, it is tentatively characterized as follows:

a. Original or superb technology and management know-how is its main asset.

b. Filled with entrepreneurial spirit devoted to business expansion.

c. Being free from domination by large companies.

In 1972–73, some venture capital companies were established, however, it was difficult for venture capital firms to realize capital gains because of the lack of the over-the-counter market, and the absence of venture capitalists handicapped the firms getting off the ground.

Fundraising is one of the biggest problems in venture business development. Venture businesses, long on future potential, are short on collateral. They generate huge capital demands for their R&D and trial production of new products.

Nevertheless, their R&D field is so advanced that the inspection and evaluation know-how has not been established, and also they do not possess sufficient material collateral. As a result, direct investment through a venture capital company is not always sufficient, and loans from banks do not always go smoothly.

Under these circumstances, Venture Enterprise Center (VEC) was established as a non-profit foundation under the civil law on July 1, 1975. It was promoted by the Ministry of International Trade & Industry, and the VEC guarantee system was established as a means of resolving this situation. Then, it forged a road for venture businesses and acted in place of venture capital.

The Center for the Development of Research And Development Oriented Enterprises in Japanese, Venture Enterprise Center in English, is a unique organization to promote the activities of venture businesses and venture capital companies in Japan. The policies devised and implemented by VEC now provide a major foundation for the promotion and cultivation of venture businesses.

Japanese venture business managed to surmount the economic stagnation resulting from the two oil crises, and the nation's "second venture business boom" eventually unfolded in about 1982, and then disappeared in 1986.

Recently, however, it seems that the third boom is appearing in Japan. The role of venture business is assuming greater and greater importance as a force for the invigoration of the national economy.

Venture businesses (VB) invigorate many kinds of creative businesses. For smooth conversion of the industrial structure, it is quite effective to develop new businesses for new industries. Venture businesses have excellent capabilities in the technological aspects that small- and medium-sized enterprises can handle; they can suggest the direction for small- and medium-sized enterprises in a knowledge-intensive industrial structure and act as a guide for small- and medium-sized enterprise management.

Also, these enterprises are achieving higher results in cooperation with other enterprises. The positive, determined entrepreneurial spirit characterized by a willingness to take risks or to challenge uncertainty could be a strong stimulus to the knowledge intensification of small- and medium-sized enterprises in general. The growth of VB will mean greater economic activity. The coming years will demand more active efforts to establish creative basic technologies and new technologies, transcending existing boundaries between industries and the expansion of their new frontiers. It is clear that venture businesses will be the key leaders of this effort and the active forces in economic activities, anywhere in the world.

The loan objectives of VEC are to act as surety for unsecured debt on loan for research and development, and to carry out information exchange by arranging study meetings, announcement meetings and so forth. It would thereby help to solve problems of fund procurement and information gathering that impede the activity of these enterprises; promote venture businesses; and contribute to the knowledge intensification of the industrial structure.

The activities of VEC can be divided into two main functions. First, to act as surety on loans from banks to venture or new service businesses; second, to carry out information exchange for the same.

## 1.1 Loan Guarantees to R&D oriented Venture Business Enterprises

## 34  Chapter Four

- A. Debt guarantee of fund loans raised by research and development oriented small- and medium-sized enterprises to finance R and D activities (mainly commercializing) for new technology and products.

- B. Qualified enterprises: Small- and medium-sized ones that aim at the development of new technology and products and possess the necessary technical and management capabilities to succeed, and have good prospects for the commercialization of their projects, under their own control.

- C. Examination: All projects applied for guarantee are examined by the examination Committee members.

- D. Limit of guarantee: 80% of loan, maximum of 100 million yen per project.

- E. Guarantee charge: 2% per year.

- F. Security: No security required beyond the quality, but the representative of the enterprise is required as guarantor.

- G. A bonus will be charged if the project is successful.

- H. Guarantee term: minimum one year, maximum eight years.

- I. Repayment procedure: Repayment is by installments.

- J. Deferment: maximum one year.

- K. Debt guarantee performance: As of the end of March 1989.

Total:  11.9 billion yen (application 36.9 billion yen)
318 projects of 276 enterprises (application 813 projects)
Success bonus: 74 million yen (19 projects)
Subrogated performance: 1,205 million yen (38 projects)
(See Table 4.3, 1).*

---

*Tables can be found at the end of the chapter.

## VECs Definition of Entrepreneurism

Entrepreneurs are motivated by certain needs in their lives or a search for the meaning of life. Spin-off entrepreneurs opted for greater freedom or had been affected by circumstances of former employment. Manufacturers seek independence of order-dominated production systems. Distributors (those handling domestic or foreign products) were affected by new needs for or by improvements in production technology in Japan. (See Table 4.3, 2, 3.)

1.2  Loan Guarantees to Knowledge-Intensive Enterprises

VEC started this new program in 1988 to foster the growth of service businesses with original and high levels of creativity on July 1, 1988.

A.  Limit of guarantee: 80% of loan, maximum of 50 million yen per project.

B.  Guarantee charge: 1% per year.

C.  Security: No security required beyond the quality, but the representative of the enterprise is required as guarantor.

D.  A bonus will be charged if the project is successful.

E.  Guarantee term: minimum one year, maximum eight years.

F.  Repayment procedure: Repayment is by installments.

G.  Deferment: maximum one year.

H.  Debt guarantee performance: As of the end of March 1989.

Total:  74 million yen; 2 Projects (2 enterprises)

36  Chapter Four

## Growth of VEC Guarantee Enterprises

The growth of 180 firms of 265 VEC first guarantee enterprises until the end of March 1988, less 42 failures and 43 unknown, is shown. (See Table 4.5.) On our recent investigation of the guarantee firms, their capital had increased by 220 percent to 21.6 billion yen from 6.8 billion yen, the employed persons by 40 percent to 14,771 from 10,755 and the annual sales by 80 percent to 340.4 billion yen from 185.8 billion yen during the years since the time of VEC guarantee, respectively.

2.1 Main information exchange activities are as follows:

    A. Organization of lectures, seminars, symposiums, to promote R&D information and management techniques exchange.

        Venture Business Management Strategy Seminars
        Venture Business Tops' Meeting
        Venture Capital Tops' Meeting
        International Symposium

    B. Guidance on management, production, sales, R&D of venture businesses.

    C. Conduct of surveys and collection, collation and processing of information regarding venture firms.

        Venture Businesses Trends Research
        Overseas Venture Enterprises Study Team

    D. Publicity for and publication of research and business activities of venture businesses.

        Venture Forum (monthly Bulletin)
        VEC Annual Report
        Patron Members lists
        Report on Venture businesses' Trends Survey

Interchange of information with the external organizations

E. Cultivation of human resources.

Training of new employees
Venture Business Tops' Seminar (risk management and marketing)

2.2 Other intermediary services for:

A. Obtaining the funds necessary for the establishment and running of R&D results.

B. The transfer of R&D results.

C. To assist in the application of technology otherwise difficult to commercialize or liable to be left unexploited by others.

D. Organization of exchanges and tie-up agreements with domestic and foreign parties relevant to R&D oriented enterprises.

3. Assets

| Fundamental Fund | 7 million yen |
| Debt Surety Fund (V.B.) | 1,200 million yen |
| Debt Surety Fund (N.S.B.) | 100 million yen |
| General Fund | 1,430 million yen |

4. Patron Members

So that VEC can hold study meetings and carry out its other information exchange activities adequately, it is trying to gain a large and extensive membership that includes venture businesses, trade, engineering and software concerns; not only small- and medium-sized enterprises but large ones as well.

4.1 Number of Patron Members: as of the end of April 1989

Small- and medium-sized business 202 companies
Venture Capital 46 companies
Bank 43 companies
Other large sized 18 companies
Foreign government and so on 13
TOTAL 322

## Activities for Venture Capital

There are over 100 venture capital companies in Japan, but they do not have yet their own association. VEC has been conducting a monthly meeting of venture capital companies since 1983, and they are occasionally studying the legal and taxation problems being faced by venture capital today.

5. VEC position in venture business fundraising for its developing stages:

   A. R&D

      Official grants, subsidies
      VECs debt guarantee loan

   B. Start-up (to develop prototype)

      Official grants, subsidies
      VECs debt guarantee loan
      Small Business Investment Company
      [Incubation centers]

   C. Start-up (commercialization)

      Official grants, subsidies
      VECs debt guarantee loan

Investment by Small Business Investment Company or
Venture capital company
[Incubation centers]

D. Early Expansion

Official grants, subsidies
Investment by Small Business Investment Company or
Venture capital company
Bank loan

E. Latter Expansion

Official grants, subsidies
Investment by Small Business Investment Company or
Venture capital company
Bank loan
Initial public offering (over-the-counter market and so on)

The Small Business Investment Companies, set up in Tokyo, Osaka, Nagoya in 1963, had invested 71.6 billion yen in stocks or convertible bonds of 1,717 small- and medium-sized firms in order to strengthen their own capital of small- and medium-sized ones, by the end of March 1989.

VEC also provides general financial assistance to small- and medium-sized enterprises and their cooperatives or associations as well as to local governments, directly or indirectly. Other financial assistance that VEC provides is as follows:

A. Grants to small- and medium-sized enterprises for their R&D projects and experimental production.

B. Financial assistance to local governments for equipment and facilities of open laboratories in public institutes.

In the above cases, R&D oriented companies are given special priority. In addition, measures are also administered across a broad range of fields:

A. The collection and supply of technical information.

B. Assistance for technical development.

C. Financing of capital for R&D.

D. Credit guarantee.

E. Support for trans-sectoral industrial cooperatives.

F. Tax incentives for experimental research spending, and more.

Venture capital refers to companies or investors who fulfill the function of supplying capital to and supporting the growth of venture businesses. Normal financial institutions find it difficult to finance these businesses, because, despite their advanced technology, they carry the excessive risk of weak management foundations.

The major operations of venture capital consist of:

A. Obtaining and ownership of stocks,

B. Obtaining and ownership of convertible bonds,

C. Providing medium- and long-term capital loans,

D. Management consulting, and so forth.

(For pool of capital, balance of investments and loans of venture capital companies, see Table 4.1, 1, 2, 3.)

Recently, while investments through venture capital have increased rapidly, loans continue to have the large portion of business. (For establishment of VC in Japan, see Table 4.1, 5.)

The limited partnership systems have been in use in Japan since April 1982. Venture capital companies act as executive partners in these funds, as well as consultants to the portfolio companies.

By the end of 1988, 73 limited partnership funds had been established in Japan with a total capitalization of 213 billion yen. (See Table 4.1, 2.)

A summary of the total investments and loans made by venture capital companies in Japan is given in Table 4.1, 1, 3.

Percentage breakdown of investments in main industries is shown in Table 4.2.

## Venture Capital Companies and VEC

Of the more than 320 VEC patron member companies, 46 are venture capitalists. Looking at these 46 companies, we should be able to understand the trends of venture business in Japan. VEC has been conducting monthly luncheon meetings of venture capital company presidents since July 1983, and nominated eight as the members of representative assembly and set up "policy board" as its subordinate in December 1984, to deal with matters related to venture capital. One of the main objectives of this meeting is to form friendships with other funds and companies.

For examples, the following problems were studied by an ad hoc group.

- A. Guidelines by Fair Trade Commission on stockholding and dispatching director to the investee enterprise. The anti-monopoly laws stipulate that a venture capital company cannot hold board seats in its portfolio companies, nor can it own more than 49 percent of the stocks of a venture business firm.

- B. Reservation for possible losses on investments

- C. Valuation of equity losses

D. Proposal to the government authorities on regulations of initial public offering of small- and medium-sized enterprises as not to suffocate them.

The relaxation of listing and registration requirements in the second section and the over-the-counter market in November 1983, sparked the second boom of establishment of venture capital companies.

In 1986, venture capital companies met with the difficulties of the investees' bankruptcy and the optimistic views and expectations were superseded by disappointment and apprehension. Nevertheless, the industry has overcome these difficulties and is now rebounding with strong confidence.

Japan's values are increasingly tending towards individualism and diversification as we approach the 21st century. Responding to this trend, venture businesses are providing the impetus for technological innovation and economic activity. And key to this is the venture capital that plays such a big role in the development of venture businesses. Compared with the United States and Europe, however, social recognition of the venture capital businesses in Japan is still insufficient due mainly to its short history.

VEC cannot give debt-guarantee to overseas venture businesses, but is able to recommend and encourage venture capital companies in Japan to invest in overseas enterprises.

Venture capital companies are to contribute much to the development of venture businesses not only in Japan but in the world. In some cases, venture capital had better act as an intermediary to combine venture business with other small or large companies as equal partners.

A venture capital company must not be so much a banker or a securities dealer but a promoter of venture businesses. VEC has added 12 venture capital companies to its list of approved financial institutions since October 1988. This action is designed not only to expand the amount of venture capitals available to venture businesses but to stimulate the growth of venture capital companies and venture investing.

In Japan, the venture capital industry's directive is to support mutual understanding, respect and cooperative reliance between venture business and venture capital companies.

*Mr. Hideo Arakawa is President of Venture Enterprise Center.*

## Table 4.1
## Venture Capital in Japan

1. VEC Member Venture Capital Companies' Balance of Investment and Loan at Year-end; unit: Million Yen

| Invest/Loan | 1982 | 1983 | 1984 | 1985 | 1986 | 1987 | 1988 | 1989 | 1990 |
|---|---|---|---|---|---|---|---|---|---|
| **Investment** | | | | | | | | | |
| 1. Stocks | 14,118 | 35,378 | 72,808 | 99,292 | 117,620 | 153,032 | 176,657 | 219,719 | 359,210 |
| Public | 1,848 | 2,131 | 3,698 | 4,597 | 9,425 | 16,825 | 20,901 | 29,875 | 54,964 |
| Private | 12,270 | 33,247 | 69,110 | 94,695 | 108,195 | 136,207 | 155,756 | 189,844 | 304,247 |
| 2. Warrant Bond | 1,209 | 8,876 | 25,832 | 45,252 | 55,081 | 62,718 | 71,307 | 99,210 | 150,029 |
| Cum-warrant | 378 | 2,013 | 5,083 | 6,840 | 7,684 | 9,334 | 9,279 | 9,454 | 12,016 |
| Ex-warrant | 831 | 6,863 | 20,749 | 38,412 | 47,397 | 53,384 | 62,028 | 89,756 | 138,012 |
| 3. Convert. Bond | 1,853 | 2,016 | 6,000 | 8,302 | 8,002 | 8,154 | 9,656 | 10,861 | 17,701 |
| 4. Total (1~3) | 17,180 | 46,270 | 104,640 | 152,846 | 180,703 | 223,904 | 257,620 | 329,790 | 526,940 |
| [Rounds]* | [361] | [628] | [1,434] | [2,055] | [2,525] | [3,035] | [3,724] | [4,215] | [5,954] |
| **Loan** | | | | | | | | | |
| 5. Bond Purchase | 2540 | 15,102 | 9,089 | 10,600 | 11,319 | 4,744 | 4,063 | 6,045 | 10,486 |
| 6. Loan | 72,360 | 103,382 | 147,916 | 142,907 | 193,314 | 301,802 | 560,863 | 614,635 | 996,125 |
| 7. Total (5, 6) | 74,900 | 118,484 | 157,005 | 153,507 | 204,633 | 306,546 | 564,926 | 620,680 | 1,006,611 |
| [Deals]* | [159] | [217] | [239] | [248] | [301] | [329] | [473] | [666] | [692] |
| Total Balance | 92,080 | 164,754 | 261,645 | 306,363 | 385,336 | 530,450 | 822,546 | 950,470 | 1,533,551 |
| [Round & Deal]* | [514] | [825] | [1,637] | [2,251] | [2,797] | [3,311] | [4,030] | [4,762] | [6,646] |
| VEC Member VCs | 9 | 24 | 32 | 36 | 39 | 40 | 42 | 43 | 47 |

Note: * The figures of total rounds or deals are avoided overlapping.

## Table 4.1 (continued)

### 2. Limited Partnership Funds Raising [L-Ps in Japan do not entirely coincide with those in U.S.A.]

| Yearly | 1982 | 1983 | 1984 | 1985 | 1986 | 1987 | 1988 | 1989 | 1990 |
|---|---|---|---|---|---|---|---|---|---|
| Established Number | 3 | 13 | 16 | 10 | 6 | 19 | 6 | 11 | 24 |
| Raised Fund (m.¥) | 5,600 | 40,450 | 46,857 | 37,200 | 15,604 | 53,389 | 13,524 | 36,171 | 115,183 |
| Partners (Investors) | 36 | 283 | 367 | 254 | 94 | 181 | 98 | 226 | 507 |
| **At Year-End** | | | | | | | | | |
| Number of Funds | 3 | 16 | 32 | 42 | 48 | 67 | 73 | 84 | 108 |
| Pool of Capital (m.¥) | 5,600 | 46,050 | 92,907 | 130,107 | 145,711 | 199,100 | 212,624 | 248,795 | 363,978 |
| Partners | 36 | 319 | 686 | 940 | 1,034 | 1,215 | 1,313 | 1,539 | 2,046 |
| Number of VC Companies managing L-P Funds | 2 | 7 | 15 | 17 | 19 | 19 | 21 | 21 | 24 |
| Investment Balance (m.¥) | 1,751 | 12,662 | 32,744 | 52,433 | 62,539 | 82,532 | 82,171 | 96,696 | 141,898 |
| Portfolio Companies | 11 | 76 | 296 | 482 | 635 | 749 | 859 | 961 | 1,311 |

### 3. Balance of Investment/Loan by Fund-Sources at Year-End; unit: Million Yen

| | 1982 | 1983 | 1984 | 1985 | 1986 | 1987 | 1988 | 1989 | 1990 |
|---|---|---|---|---|---|---|---|---|---|
| Investment | | | | | | | | | |
| Own Capital | 15,429 | 33,608 | 71,896 | 100,413 | 118,164 | 141,372 | 175,449 | 233,094 | 385,042 |
| L-P Fund | 1,751 | 12,662 | 32,744 | 52,433 | 62,539 | 82,532 | 82,171 | 96,696 | 141,898 |
| Total | 17,180 | 46,270 | 104,640 | 152,846 | 180,703 | 223,904 | 257,620 | 329,790 | 526,940 |
| Loan Own Cap. | 74,900 | 118,484 | 157,005 | 153,507 | 204,633 | 306,546 | 564,926 | 620,680 | 1,006,611 |
| Reserved L-P Fund | 3,849 | 33,388 | 60,163 | 77,674 | 83,172 | 116,568 | 130,453 | 152,099 | 222,080 |
| Available | 95,929 | 198,142 | 321,808 | 384,027 | 468,508 | 647,018 | 952,999 | 1,102,569 | 1,755,631 |

## Table 4.1 (continued)

### 4. Percentage Breakdown of Year-End Investments in Main 12 Industries; Own=Own Capital

| Industry/Fund | 1982 Own | 1982 L-P | 1983 Own | 1983 L-P | 1985 Own | 1985 L-P | 1986 Own | 1986 L-P | 1987 Own | 1987 L-P | 1988 Own | 1988 L-P | 1989 Own | 1989 L-P | 1990 Own | 1990 L-P |
|---|---|---|---|---|---|---|---|---|---|---|---|---|---|---|---|---|
| Commercial | $13^9$ | $39^4$ | $10^3$ | $10^6$ | $10^1$ | $5^7$ | $10^9$ | $6^7$ | $9^7$ | $8^2$ | $12$ | $8^8$ | $14^6$ | $11^8$ | $16$ | $13^8$ |
| Financing | | | | | $6^3$ | $3^6$ | $8^3$ | $3^8$ | $9^4$ | $4^6$ | $9^2$ | $4^8$ | $8^1$ | $5^6$ | $12^5$ | $6^6$ |
| Electric equip. | $7^8$ | 0 | $11^8$ | $10^9$ | $23^7$ | $27^9$ | $22^6$ | $26^3$ | $21^4$ | $25^5$ | $16^1$ | $22^1$ | $11^4$ | $19^3$ | $8^7$ | $14^6$ |
| Inf. service | $13^8$ | 0 | $15^2$ | $4^3$ | $7^5$ | $4^8$ | $7^5$ | $5^2$ | $6^4$ | $4^6$ | $6^4$ | $5^1$ | $9^5$ | $7^3$ | $8^3$ | $7^9$ |
| Foreign company | 0 | 0 | 0 | 0 | $2^8$ | 6 | $3^2$ | $15^6$ | $4^1$ | $13^9$ | $4^8$ | $13^4$ | $2^2$ | $13^1$ | $4^9$ | $9^3$ |
| General equip. | $8^5$ | $13^2$ | 7 | $10^3$ | $5^5$ | $3^6$ | $8^1$ | $5^3$ | $6^9$ | 6 | $7^6$ | $7^1$ | $6^3$ | 7 | $5^9$ | $7^6$ |
| Construction | | | | | $1^7$ | $2^4$ | $1^7$ | $2^3$ | $2^2$ | $2^7$ | $3^2$ | $4^3$ | $4^5$ | 4 | $5^8$ | 5 |
| Other mfg. | $6^8$ | 10 | $11^1$ | $28^4$ | $6^2$ | $6^7$ | 6 | $4^2$ | $5^5$ | $3^7$ | $5^7$ | $3^6$ | $6^4$ | $3^4$ | $5^8$ | $4^7$ |
| Chemical | $5^4$ | 0 | $6^8$ | $1^8$ | $5^8$ | $2^5$ | $4^6$ | $1^9$ | $5^1$ | $2^8$ | $4^8$ | $2^7$ | $4^9$ | $3^5$ | $4^7$ | $3^3$ |
| Software | $0^1$ | 0 | $2^1$ | $1^1$ | $4^7$ | $7^9$ | $5^9$ | $7^8$ | $5^8$ | $6^2$ | $5^5$ | $6^6$ | $4^2$ | $4^6$ | $3^8$ | $4^5$ |
| Real estate | | | | | $0^9$ | $2^8$ | $1^2$ | $0^2$ | $2^8$ | $1^8$ | 2 | $3^1$ | $2^5$ | $1^9$ | $3^8$ | $4^2$ |
| Precision m/c. | $3^9$ | $20^1$ | $4^6$ | $9^9$ | $7^8$ | $8^4$ | $5^8$ | $6^1$ | $5^5$ | 6 | $6^9$ | $5^7$ | $5^2$ | $4^6$ | $3^5$ | 4 |
| Sub-total | $60^2$ | $82^7$ | $68^9$ | $77^3$ | $83^0$ | $82^3$ | $85^8$ | $85^4$ | $84^8$ | $86^0$ | $84^2$ | $87^3$ | $79^8$ | $86^1$ | $83^7$ | $85^5$ |
| Total | 100 | 100 | 100 | 100 | 100 | 100 | 100 | 100 | 100 | 100 | 100 | 100 | 100 | 100 | 100 | 100 |
| Invested (b.¥) | 15.4 | 1.8 | 33.6 | 12.6 | 100.4 | 52.4 | 118.2 | 62.5 | 141.4 | 82.5 | 175.4 | 82.2 | 233.1 | 96.7 | 385.0 | 141.9 |
| VC | 9 | 2 | 22 | 7 | 33 | 17 | 36 | 19 | 37 | 19 | 39 | 21 | 41 | 21 | 47 | 24 |

Note: Figures of 1984 are not available.

## Table 4.1 (continued)

5. Establishment of Venture Capital Companies in Japan; Number of companies, total capital and affiliations. As of the beginning of March of 1991. Unit of Capital: Million Yen, and shows up-to-date capital. Number in ( ) shows VEC members' among them.

| Year | Affiliates of Banks No. | Capital | Affiliates of Securities Corp. No. | Capital | Independent No. | Capital | Affiliates of Foreign Corp. No. | Capital | Total No. | Capital |
|---|---|---|---|---|---|---|---|---|---|---|
| 1972 | 1 (1) | 4,400 (4,400) | — | — | — | — | — | — | 1 (1) | 4,400 (4,400) |
| 1973 | — | — | 2 (2) | 27,462 (27,462) | — | — | — | — | 2 (2) | 27,462 (27,462) |
| 1974 | 3 (3) | 1,950 (1,950) | — | — | 1 (1) | 500 (500) | — | — | 4 (4) | 2,450 (2,450) |
| 1980 | — | — | — | — | — | — | 1 (–) | 40 (–) | 1 (–) | 40 (–) |
| 1982 | — | — | 6 (5) | 3,618 (3,418) | 1 (–) | 100 (–) | 1 (–) | 6 (–) | 8 (5) | 3,724 (3,418) |
| 1983 | 2 (1) | 550 (450) | 7 (7) | 2,320 (2,320) | 5 (2) | 420 (320) | — | — | 14 (10) | 3,290 (3,090) |
| 1984 | 22 (9) | 3,530 (3,030) | 1 (1) | 320 (320) | 3 (1) | 74 (10) | 1 (–) | 30 (–) | 27 (11) | 3,945 (3,360) |
| 1985 | 15 (5) | 1,550 (720) | 2 (1) | 700 (600) | 2 (–) | 60 (–) | 1 (1) | 5 (5) | 20 (7) | 2,315 (1,325) |
| 1986 | 4 (2) | 520 (400) | — | — | 1 (–) | 90 (–) | — | — | 5 (2) | 610 (400) |
| 1987 | — | — | 1 (1) | 300 (300) | — | — | — | — | 1 (1) | 300 (300) |
| 1988 | 1 (1) | 500 (500) | — | — | 2 (–) | 60 (–) | — | — | 3 (1) | 560 (500) |
| 1989 | 3 (1) | 550 (450) | 1 (1) | 300 (300) | — | — | — | — | 4 (2) | 850 (750) |
| 1990 | 7 (2) | 2,580 (680) | 2 (–) | 300 (–) | 4 (–) | 2,850 (–) | — | — | 13 (2) | 5,730 (680) |
| 1991 | 1 (–) | 100 (–) | — | — | — | — | — | — | 1 (–) | 100 (–) |
| Total | 59 (25) | 16,230 (12,580) | 22 (18) | 35,320 (34,720) | 19 (4) | 4,154 (830) | 4 (1) | 81 (5) | 104 (48) | 55,785 (48,135) |

Note: Following organizations are not included:
VEC
Small Business Investment Companies (Tokyo, Nagoya, Osaka) and the affiliated Fund of them.
Venture Business Development Foundation (SanwaBank, Fukuoka Bank, Hokutaku Bank, Osaka Prefecture)
New Business Investment Ltd., Co.
Japan Asean Investment Ltd., Co. and some others.

## Table 4.2
### Percentage Breakdown of Venture Capital Investments by Industry Category

| at the end of | 1982 Own | 1982 L-P | 1983 Own | 1983 L-P | 1985 Own | 1985 L-P | 1986 Own | 1986 L-P | 1987 Own | 1987 L-P | 1988 Own | 1988 L-P | 1989 Own | 1989 L-P | 1990 Own | 1990 L-P |
|---|---|---|---|---|---|---|---|---|---|---|---|---|---|---|---|---|
| Industry | | | | | | | | | | | | | | | | |
| Food mfg. | 1.8 | 0 | 1.3 | 1.5 | 3.0 | 4.6 | 1.7 | 2.5 | 2.5 | 3.1 | 2.6 | 4.2 | 3.5 | 3.4 | 2.6 | 3.2 |
| Textile | 0.7 | 0 | 0.8 | 0 | 0.2 | 0 | 0 | 0 | 0.2 | 0 | 0.1 | 0 | 0.7 | 0 | 0 | 0 |
| Clothing | 2.9 | 10.0 | 1.6 | 2.3 | 0.8 | 1.1 | 0.4 | 0.7 | 0.3 | 0.5 | 0.4 | 0.4 | 0.5 | 0.8 | 1.0 | 0.9 |
| Lumber | 0.2 | 0 | 0.1 | 0 | 0.2 | 0.7 | 0.1 | 0 | 0.2 | 0.2 | 0.3 | 0.6 | 0.3 | 0.4 | 0.1 | 0.2 |
| Furniture | 0.7 | 0 | 0.5 | 0.3 | 1.0 | 0.3 | 0.6 | 0.5 | 0.7 | 0.5 | 0.7 | 0.8 | 0.6 | 0.5 | 0.5 | 0.6 |
| Pulp, paper | 0 | 0 | 0.1 | 0 | 0.9 | 0.7 | 0.3 | 0 | 0.4 | 0.2 | 0.1 | 0.2 | 0.3 | 0.2 | 0.6 | 0.8 |
| Printing | 0.3 | 0 | 0.2 | 0.4 | 0.6 | 0.5 | 0.6 | 0.4 | 1.0 | 0.5 | 0.9 | 0.5 | 1.1 | 0.4 | 1.6 | 1.5 |
| Chemical | 5.4 | 0 | 6.8 | 1.8 | 5.8 | 2.5 | 4.6 | 2.9 | 5.1 | 2.8 | 5.0 | 2.2 | 4.9 | 3.5 | 4.7 | 3.3 |
| Oil & coal | 0.1 | 0 | 0.1 | 0 | 0.1 | 0.3 | 0.1 | 0 | 0.1 | 0 | 0.1 | 0 | 0.1 | 0 | 0.2 | 0 |
| Plastics | unknown | | unknown | | unknown | | 0.4 | 1.0 | 0.4 | 0.7 | 0.5 | 0.5 | 0.8 | 1.1 | 1.0 | 1.0 |
| Rubber | 0.7 | 0 | 0.3 | 0 | 0 | 0.4 | 0.3 | 0 | 0.2 | 0 | 0.5 | 0.1 | 1.2 | 0.1 | 0.3 | 0.1 |
| Ceramics | 0.1 | 0 | 0.5 | 0 | 1.3 | 1.0 | 1.2 | 0.6 | 1.0 | 0.7 | 0.7 | 0.6 | 0.7 | 0.7 | 0.7 | 0.7 |
| Steel | 0.1 | 0 | 0.1 | 0 | 0 | 0 | 0 | 0 | 0 | 0 | 0.1 | 0 | 0.3 | 0 | 0.5 | 0.1 |
| Non-ferrous | 0 | 0 | 0 | 0 | 0.4 | 0.3 | 0.2 | 0.2 | 0.1 | 0.1 | 0.2 | 0.2 | 0.3 | 0.3 | 0.1 | 0.3 |
| Metal | 1.6 | 0 | 1.0 | 0 | 1.7 | 1.7 | 1.5 | 0.9 | 1.4 | 0.9 | 1.5 | 1.1 | 1.6 | 0.8 | 2.0 | 1.8 |
| General equip. | 8.5 | 13.2 | 7.0 | 10.3 | 5.5 | 3.6 | 8.1 | 5.3 | 6.9 | 6.0 | 7.5 | 7.8 | 6.3 | 7.0 | 5.9 | 7.6 |
| Electric m/c. | 7.8 | 0 | 11.8 | 10.9 | 23.7 | 27.9 | 22.6 | 26.6 | 21.4 | 25.5 | 15.9 | 25.0 | 11.4 | 19.3 | 8.7 | 14.6 |
| Transport equip. | 1.9 | 0 | 1.0 | 0.6 | 1.0 | 0.4 | 1.2 | 0.5 | 1.4 | 0.4 | 1.1 | 0.3 | 1.3 | 0.2 | 1.1 | 0.6 |
| Precision m/c. | 3.9 | 20.1 | 4.6 | 9.9 | 7.8 | 8.4 | 5.8 | 6.1 | 5.5 | 6.0 | 7.5 | 4.5 | 5.2 | 4.6 | 3.5 | 4.0 |
| Other mfg. | 6.8 | 10.0 | 11.1 | 28.4 | 6.2 | 6.7 | 6.0 | 4.2 | 5.5 | 3.7 | 5.5 | 4.3 | 6.4 | 3.4 | 5.8 | 4.7 |

## Table 4.2 (continued)

| at the end of | 1982 Own | 1982 L-P | 1983 Own | 1983 L-P | 1985 Own | 1985 L-P | 1986 Own | 1986 L-P | 1987 Own | 1987 L-P | 1988 Own | 1988 L-P | 1989 Own | 1989 L-P | 1990 Own | 1990 L-P |
|---|---|---|---|---|---|---|---|---|---|---|---|---|---|---|---|---|
| Industry |  |  |  |  |  |  |  |  |  |  |  |  |  |  |  |  |
| Software | 0.1 | 0 | 2.1 | 1.1 | 4.7 | 7.9 | 5.9 | 7.8 | 5.8 | 6.2 | 5.5 | 6.7 | 4.2 | 4.6 | 3.8 | 4.5 |
| Inf. process. | 0.3 | 0 | 1.2 | 4.3 | 2.3 | 2.8 | 2.0 | 1.0 | 1.7 | 0.5 | 1.2 | 0.8 | 0.7 | 0.3 | 0.8 | 0.5 |
| Inf. supply svc. | 0.3 | 0 | 0.2 | 0 | 0.2 | 0.6 | 0.4 | 0.9 | 0.4 | 0.7 | 0.4 | 0.5 | 0.6 | 0.5 | 0.2 | 0.4 |
| Other inf. svc. | 3.0 | 0 | 3.1 | 0 | 0.8 | 0.3 | 0.8 | 0.7 | 0.4 | 0.3 | 0.4 | 0.5 | 0.2 | 0.5 | 0.1 | 0 |
| Other svc. | 10.2 | 0 | 10.7 | 0 | 4.2 | 1.1 | 4.3 | 2.6 | 3.9 | 3.1 | 4.4 | 4.2 | 8.0 | 6.0 | 7.2 | 7.0 |
| Commercial | 13.9 | 39.4 | 10.3 | 10.6 | 10.1 | 5.7 | 10.9 | 6.7 | 9.7 | 8.2 | 11.5 | 14.0 | 14.6 | 11.8 | 16.0 | 13.8 |
| Others | 28.7 | 7.3 | 23.5 | 17.6 | 14.8 | 14.5 | 16.9 | 12.3 | 19.7 | 15.3 | 20.5 | 18.1 | 22.6 | 16.6 | 25.7 | 18.8 |
| Restaurant |  |  |  |  | 1.2 | 0.3 | 1.6 | 0.7 | 1.6 | 0.9 | 1.7 | 1.7 | 3.4 | 2.4 | 1.4 | 1.3 |
| Construction |  |  |  |  | 1.7 | 2.4 | 1.7 | 2.3 | 2.2 | 2.7 | 3.2 | 4.3 | 4.5 | 4.0 | 5.8 | 5.0 |
| Financing | 18.9 |  | 17.2 |  | 6.3 | 3.6 | 8.3 | 3.8 | 9.4 | 4.6 | 9.2 | 4.8 | 8.1 | 5.6 | 12.5 | 6.6 |
| Transport |  |  |  |  | 0.2 | 0.2 | 0.2 | 0.1 | 0.2 | 0 | 0.3 | 0 | 0.8 | 0.1 | 0.6 | 0.2 |
| Real estate |  |  |  |  | 0.9 | 2.8 | 1.2 | 0.2 | 2.8 | 1.8 | 2.0 | 3.1 | 2.5 | 1.9 | 3.8 | 4.2 |
| Trading co. |  |  |  |  | 2.3 | 4.2 | 2.8 | 4.1 | 2.9 | 4.2 | 3.4 | 4.1 | 1.5 | 2.0 | 1.0 | 0.9 |
| Others |  |  |  |  | 2.2 | 1.1 | 1.1 | 1.1 | 0.6 | 1.1 | 0.7 | 0.1 | 1.8 | 0.6 | 0.6 | 0.6 |
| Foreign co. |  |  |  |  | 2.8 | 6.0 | 3.2 | 15.6 | 4.1 | 13.9 | 5.0 | 5.6 | 2.2 | 13.1 | 4.9 | 9.3 |
| Total | 100.0 | 100.0 | 100.0 | 100.0 | 100.0 | 100.0 | 100.0 | 100.0 | 100.0 | 100.0 | 100.0 | 100.0 | 100.0 | 100.0 | 100.0 | 100.0 |
| Invested (billion Yen) | 15.4 | 1.8 | 33.6 | 12.7 | 100.4 | 52.4 | 118.2 | 62.5 | 141.4 | 82.5 | 175.4 | 82.2 | 233.1 | 96.7 | 385.0 | 141.9 |
| No. of VC companies | 9 | 2 | 22 | 7 | 33 | 17 | 36 | 19 | 37 | 19 | 39 | 21 | 43 | 21 | 47 | 24 |

Note: Figures of 1984 are not available
OWN = Own Capital
L-P = Limited Partnership Fund

## Table 4.3
## Some Findings of VEC Guarantee Enterprises (so-called Venture Business)

1. Breakdown of VEC Guarantees at the time of Guarantee by Commodity Category (as of the end of March 1991—cumulative totals)

remarks: Pr: project, perf; performance, A; at the time of subrogation, B; at the time of guarantee

| Commodity to be developed | Enter-prises | Capital m.Yen | Avg. | Em-ployed persons | Avg. | Guaran-tee m.Yen | Avg. | Success Pr. | Bonus m.Yen | Subrogated perf. Pr. | A mYen | B mYen |
|---|---|---|---|---|---|---|---|---|---|---|---|---|
| Energy Equip. etc. | 6 | 83 | 13.8 | 97 | 16.2 | 172.8 | 28.8 | — | — | 1 | 3.22 | 7.2 |
| Machine elements | 6 | 148 | 24.7 | 494 | 82.3 | 225.6 | 37.6 | — | — | 3 | 67.15 | 85.6 |
| Machine tools | 19 | 987 | 51.9 | 1,273 | 67.0 | 655.2 | 34.5 | 1 | 4.0 | 2 | 78.45 | 84.0 |
| Power equip. | 4 | 124 | 31.0 | 166 | 41.5 | 216.0 | 54.0 | — | — | 1 | 67.50 | 80.0 |
| Equip. for chemical industries | 3 | 187 | 62.3 | 136 | 45.3 | 142.4 | 47.5 | — | — | 1 | 42.94 | 48.0 |
| Industrial equip. | 48 | 1,890 | 39.4 | 2,588 | 53.9 | 1,934.7 | 40.3 | 6 | 30.4 | 4 | 96.53 | 164.8 |
| Scientific equip. | 54 | 2,489 | 46.1 | 2,616 | 48.4 | 1,926.4 | 35.7 | 2 | 4.0 | 4 | 99.72 | 128.0 |
| Office equip. etc. | 8 | 332 | 41.5 | 381 | 47.6 | 357.6 | 44.7 | 1 | 4.0 | 3 | 131.37 | 156.0 |
| Elect., commun. equip. | 21 | 1,007 | 48.0 | 1,029 | 49.0 | 738.4 | 35.2 | — | — | 4 | 125.72 | 178.4 |
| Peripheral equip. of computer | 48 | 1,585 | 33.0 | 1,758 | 36.6 | 1,874.4 | 39.1 | 4 | 10.6 | 3 | 111.93 | 152.0 |

## Table 4.3 (continued)

| Commodity to be developed | Enter-prises | Capital m.Yen | Avg. | Em-ployed persons | Avg. | Guaran-tee m.Yen | Avg. | Success Pr. | Bonus m.Yen | Subrogated perf. Pr. | A mYen | B mYen |
|---|---|---|---|---|---|---|---|---|---|---|---|---|
| Elect. appl. equip. | 50 | 1,676 | 33.5 | 2,637 | 52.7 | 2,267.2 | 45.3 | 6 | 26.6 | 7 | 210.49 | 374.4 |
| Transport. equip. | 7 | 121 | 17.3 | 254 | 36.3 | 229.6 | 32.8 | 2 | 7.5 | 1 | 42.72 | 40.0 |
| Anti-pollution etc. | 30 | 774 | 25.8 | 1,695 | 56.5 | 1,069.6 | 35.7 | — | — | 7 | 246.20 | 352.0 |
| Ceramics | 9 | 682 | 75.8 | 1,120 | 124.4 | 506.4 | 56.3 | — | — | — | — | — |
| Iron & steel | 5 | 482 | 96.4 | 207 | 41.4 | 264.0 | 52.8 | — | — | 1 | 21.08 | 48.0 |
| Non-ferrous product | 7 | 124 | 17.7 | 185 | 26.4 | 243.2 | 34.7 | — | — | — | — | — |
| Wires, cables, etc. | 1 | 40 | 40.0 | 4 | 4.0 | 44.0 | 44.0 | — | — | — | — | — |
| Chemical products | 17 | 370 | 21.8 | 875 | 51.5 | 820.8 | 48.3 | — | — | 1 | 33.14 | 36.0 |
| Synthetic resins | 6 | 141 | 23.5 | 488 | 81.3 | 160.0 | 26.7 | — | — | 1 | 11.30 | 12.0 |
| Textile, paper, pulp | — | — | — | — | — | — | — | — | — | — | — | — |
| Construction etc. | 11 | 732 | 66.5 | 745 | 67.7 | 396.8 | 36.1 | — | — | 1 | 19.54 | 40.0 |
| Living goods | 4 | 128 | 32.0 | 113 | 28.3 | 160.0 | 40.0 | — | — | — | — | — |
| Total | 364 | 14,102 | 38.7 | 18,861 | 51.8 | 14,405.1 | 39.6 | 22 | 87.1 | 45 | 1,409.00 | 1,986.4 |

Note: Figures are cumulative totals as of the end of March 1991.
Ratio of subrogated project number (45) to the total (364) is 12.4%, and ratio of sum at the guarantee (1,986 million yen) to the total (14,405.1 million yen) is 13.8%.
Applications were made for 863 projects and 40,342 million yen.

Guarantee to knowledge-intensive enterprise: 181 million yen for 7 projects (application for 20 projects and 700 million yen).

## Table 4.4
## Some Findings of VEC Guarantee Enterprises (Venture Business)

### 1. Facts on Presidents of Enterprises

Although the total number of guaranteed projects is 364, the actual number of guaranteed enterprises is 314, as some of them have been guaranteed by VEC for their two or more projects.

#### 1. Age of Presidents

| Age | ~29 | 30~34 | 35~39 | 40~44 | 45~49 | 50~59 | 60~69 | 70~ | no fou | unknown | Total |
|---|---|---|---|---|---|---|---|---|---|---|---|
| at Est | 73 | 43 | 39 | 49 | 20 | 24 | 1 | 1 | 63 | 1 | 314 p |
| ditto % | 23.2 | 13.7 | 12.4 | 15.6 | 6.4 | 7.6 | 0.3 | 0.3 | 20.1 | 0.3 | 100.0 |
| at Gua | 5 | 12 | 37 | 55 | 57 | 116 | 25 | 6 | — | 1 | 314 p |
| ditto % | 1.6 | 3.8 | 11.8 | 17.5 | 18.2 | 36.9 | 8.0 | 1.9 | — | 0.3 | 100.0 |

Remarks: at Est = at the time of establishment; at Gua = at the time of guarantee; no fou = not founder; p = persons

#### 2. President's business career background

| | persons | % |
|---|---|---|
| None | 51 | 16.2 |
| From large enterprises | 90 | 28.7 |
| From large enterprise through small and medium enterprise | 33 | 10.5 |
| From small and medium ent. | 123 | 39.2 |
| From public service | 14 | 4.5 |
| Unknown | 3 | 0.9 |
| Total | 314 | 100.0 |

## Table 4.4 (continued)

### 3. President's acedemic background

| | persons | % |
|---|---|---|
| Junior high school graduate | 7 | 2.2 |
| Senior high school graduate | 48 | 15.3 |
| Junior college etc. graduate | 33 | 10.5 |
| University (Art) graduate | 72 | 22.9 |
| University (Sciences) graduate | 114 | 36.3 |
| Graduate school (Sciences) | 14 | 4.5 |
| Others | 22 | 7.0 |
| Unknown | 4 | 1.3 |
| Total | 314 | 100.0 |

## 2. Facts on Enterprises

### 1. Capital at the time of guarantee

| | company | % |
|---|---|---|
| Less than 10 million Yen | 70 | 19.2 |
| 10 – 30 million Yen | 144 | 39.6 |
| 30 – 50 million Yen | 75 | 20.6 |
| 50 – 100 million Yen | 47 | 12.9 |
| Over 100 million Yen | 28 | 7.7 |
| Total | 364 | 100.0 |

## Table 4.4 (continued)

### 2. Sales at the time of guarantee

| | company | % |
|---|---|---|
| Less than 50 million Yen | 29 | 8.0 |
| 50 – 100 million Yen | 20 | 5.5 |
| 100 – 300 million Yen | 79 | 21.7 |
| 300 – 500 million Yen | 55 | 15.1 |
| 0.5 – 1 billion Yen | 78 | 21.4 |
| 1 – 2 billion Yen | 55 | 15.1 |
| Over 2 billion Yen | 48 | 13.2 |
| Total | 364 | 100.0 |

### 3. Number of employees at the time of guarantee

| | company | % |
|---|---|---|
| 4 and less persons | 18 | 4.9 |
| 5 – 19 persons | 138 | 37.9 |
| 20 – 49 persons | 96 | 26.4 |
| 50 – 99 persons | 56 | 15.4 |
| 100 – 299 persons | 51 | 14.0 |
| Over 300 persons | 5 | 1.4 |
| Total | 364 | 100.0 |

## Table 4.4 (continued)

*4. Years after establishment at the time of guarantee*

| | company | % |
|---|---|---|
| Less than 3 years | 51 | 16.2 |
| 3 – 5 years | 36 | 11.5 |
| 5 – 7 years | 32 | 10.2 |
| 7 – 10 years | 44 | 14.0 |
| 10 – 15 years | 51 | 16.2 |
| 15 – 20 years | 26 | 8.3 |
| 20 – 30 years | 47 | 15.0 |
| Over 30 years | 27 | 8.6 |
| Total | 314 | 100.0 |

## Table 4.5
## Growth of VEC Guarantee Enterprises by Yearly Groups

### 1. Growth of Capital, Employed Persons and Annual Sales

| Fiscal Year | Enter-prises | Capital A Bil.y | Capital B Bil.y | Capital B/A | Employed A x100 | Employed B x100 | Employed B/A | Annual Sales A Bil.y | Annual Sales B Bil.y | Annual Sales B/A | No. of Enterprises Total Guar. | No. of Enterprises 1st Guar. | No. of Enterprises Subrogated | No. of Enterprises Unknown |
|---|---|---|---|---|---|---|---|---|---|---|---|---|---|---|
| 1975 | 22 | .95 | 3.36 | 3.5 | 15.9 | 26.7 | 1.7 | 17.76 | 83.38 | 4.7 | 39 | 39 | 9 | 8 |
| 1976 | 21 | 1.18 | 8.55 | 7.2 | 19.7 | 29.7 | 1.5 | 29.78 | 83.19 | 2.8 | 34 | 34 | 1 | 12 |
| 1977 | 6 | .08 | 2.02 | 24.6 | 2.3 | 4.8 | 2.1 | 2.02 | 16.88 | 8.4 | 11 | 11 | 4 | 1 |
| 1978 | 4 | .12 | .63 | 5.2 | 1.8 | 2.7 | 1.5 | 2.30 | 5.35 | 2.3 | 10 | 5 | 0 | 1 |
| 1979 | 5 | .16 | 1.62 | 10.0 | 2.5 | 7.7 | 3.0 | 5.13 | 17.34 | 3.4 | 12 | 10 | 1 | 4 |
| 1980 | 2 | .02 | .05 | 2.2 | 0.4 | 1.2 | 2.9 | .71 | 2.40 | 3.4 | 10 | 4 | 0 | 2 |
| 1981 | 9 | .17 | .72 | 4.2 | 2.6 | 4.4 | 1.7 | 4.74 | 8.86 | 1.9 | 18 | 14 | 3 | 2 |
| 1982 | 13 | .29 | 1.28 | 4.4 | 12.9 | 16.6 | 1.3 | 22.97 | 37.87 | 1.6 | 33 | 23 | 3 | 7 |
| 1983 | 21 | .74 | 3.99 | 5.4 | 12.0 | 16.6 | 1.4 | 30.95 | 42.25 | 1.4 | 35 | 33 | 7 | 5 |
| 1984 | 23 | 1.01 | 2.51 | 2.5 | 15.1 | 18.7 | 1.2 | 27.80 | 35.69 | 1.3 | 36 | 34 | 7 | 4 |
| 1985 | 21 | .58 | 1.90 | 3.3 | 8.6 | 12.7 | 1.5 | 15.72 | 25.75 | 1.6 | 32 | 30 | 4 | 5 |
| 1986 | 13 | .43 | .65 | 1.5 | 9.1 | 9.4 | 1.0 | 16.60 | 19.84 | 1.2 | 22 | 16 | 2 | 1 |
| 1987 | 11 | .48 | .62 | 1.3 | 5.1 | 4.9 | 1.0 | 8.32 | 10.46 | 1.3 | 14 | 12 | 1 | 0 |
| 1988 | 11 | .95 | .99 | 1.0 | 7.3 | 8.2 | 1.1 | 15.68 | 19.29 | 1.2 | 12 | 11 | 0 | 0 |
| Total | 182 | 7.17 | 28.87 | 4.0 | 115.5 | 164.3 | 1.4 | 200.48 | 408.53 | 2.0 | 318 | 276 | 42 | 52 |

Note: A; at the time of guarantee
B; at present
Unknown and so on; includes bankrupts except subrogated

## Table 4.5 (continued)

### 2. Growth of Capital Turnover, Sales and Capital per an Employee

| Fiscal Year | Enter-prises | Annual Sales/Capital (times) A | B | (%) B/A | Annual Sales/Employee (million Yen) A | B | (%) B/A | Capital/Employee (million Yen) A | B | (%) B/A |
|---|---|---|---|---|---|---|---|---|---|---|
| 1975 | 22 | 18.7 | 24.8 | 132.6 | 11.2 | 31.2 | 278.6 | 0.6 | 1.3 | 216.7 |
| 1976 | 21 | 25.2 | 9.7 | 38.5 | 15.1 | 28.0 | 185.4 | 0.6 | 2.9 | 483.3 |
| 1977 | 6 | 24.6 | 8.7 | 35.4 | 8.6 | 35.2 | 409.3 | 0.4 | 4.2 | 1,050.0 |
| 1978 | 4 | 19.2 | 8.5 | 44.3 | 12.5 | 20.0 | 160.0 | 0.6 | 2.3 | 383.3 |
| 1979 | 5 | 31.8 | 10.7 | 33.6 | 20.2 | 22.5 | 111.4 | 0.6 | 2.1 | 350.0 |
| 1980 | 2 | 32.3 | 50.0 | 154.8 | 17.3 | 20.0 | 115.6 | 0.5 | 0.4 | 80.0 |
| 1981 | 9 | 27.9 | 12.3 | 44.1 | 18.0 | 20.1 | 111.7 | 0.6 | 1.6 | 266.6 |
| 1982 | 13 | 79.2 | 29.5 | 37.2 | 17.9 | 22.9 | 127.9 | 0.2 | 0.8 | 400.0 |
| 1983 | 21 | 41.9 | 10.6 | 25.3 | 25.7 | 25.5 | 99.2 | 0.6 | 2.4 | 400.0 |
| 1984 | 23 | 27.4 | 14.2 | 51.8 | 18.4 | 19.1 | 103.8 | 0.7 | 1.3 | 185.7 |
| 1985 | 21 | 27.2 | 13.6 | 50.0 | 18.3 | 20.3 | 110.9 | 0.7 | 1.5 | 214.3 |
| 1986 | 13 | 38.5 | 30.4 | 79.0 | 18.2 | 21.1 | 115.9 | 0.5 | 0.7 | 140.0 |
| 1987 | 11 | 17.3 | 17.0 | 98.3 | 16.3 | 21.4 | 131.3 | 0.9 | 1.3 | 144.4 |
| 1988 | 11 | 16.6 | 19.6 | 118.1 | 21.4 | 23.4 | 109.3 | 1.3 | 1.2 | 92.3 |
| Total | 182 | 28.0 | 14.1 | 50.4 | 17.4 | 24.9 | 143.1 | 0.6 | 1.8 | 300.0 |

## Table 4.5 (continued)

### 3. Average Growth per VEC Guarantee Enterprise

| Fiscal Year | Enterprises | at the time of Guarantee ||| at present |||
|---|---|---|---|---|---|---|---|
| | | Capital m.y | Employed pers. | Annual Sales m.y | Capital m.y | Employed pers. | Annual Sales m.y |
| 1975 | 22 | 43.3 | 72.3 | 807.3 | 152.9 | 121.5 | 3,789.8 |
| 1976 | 21 | 56.2 | 93.6 | 1,418.1 | 406.9 | 141.5 | 3,961.5 |
| 1977 | 6 | 13.7 | 39.0 | 336.7 | 336.2 | 80.0 | 2,812.5 |
| 1978 | 4 | 29.9 | 46.0 | 575.0 | 156.5 | 66.8 | 1,337.5 |
| 1979 | 5 | 32.3 | 50.8 | 1,026.4 | 323.5 | 153.8 | 3,467.8 |
| 1980 | 2 | 11.0 | 20.5 | 355.0 | 24.0 | 60.0 | 1,200.0 |
| 1981 | 9 | 18.8 | 29.2 | 526.1 | 80.0 | 48.9 | 984.6 |
| 1982 | 13 | 22.3 | 98.8 | 1,767.2 | 98.6 | 127.5 | 2,912.8 |
| 1983 | 21 | 35.2 | 57.3 | 1,474.0 | 190.0 | 79.0 | 2,012.0 |
| 1984 | 23 | 44.0 | 65.7 | 1,208.5 | 109.1 | 81.4 | 1,551.6 |
| 1985 | 21 | 27.6 | 40.8 | 748.4 | 90.4 | 60.3 | 1,226.0 |
| 1986 | 13 | 33.2 | 70.3 | 1,277.0 | 50.2 | 72.2 | 1,526.2 |
| 1987 | 11 | 43.8 | 46.5 | 756.1 | 56.0 | 44.4 | 951.1 |
| 1988 | 11 | 85.9 | 66.7 | 1,425.7 | 89.6 | 74.8 | 1,753.2 |
| Total | 182 | 39.4 | 63.4 | 1,101.5 | 158.6 | 90.3 | 2,244.7 |

## Table 4.5 (continued)

### 4. Index Number (total average is one)

| Fiscal Year | at the time of Guarantee ||| at present |||
|---|---|---|---|---|---|---|
| | Capital | Employed | Annual Sales | Capital | Employed | Annual Sales |
| 1975 | 1.1 | 1.1 | 0.7 | 1.0 | 1.3 | 1.7 |
| 1976 | 1.4 | 1.5 | 1.3 | 2.6 | 1.6 | 1.8 |
| 1977 | 0.3 | 0.6 | 0.3 | 2.1 | 0.9 | 1.3 |
| 1978 | 0.8 | 0.7 | 0.5 | 1.0 | 0.7 | 0.6 |
| 1979 | 0.8 | 0.8 | 0.9 | 2.0 | 1.7 | 1.5 |
| 1980 | 0.3 | 0.3 | 0.3 | 0.2 | 0.7 | 0.5 |
| 1981 | 0.5 | 0.5 | 0.5 | 0.5 | 0.5 | 0.4 |
| 1982 | 0.6 | 1.6 | 1.6 | 0.6 | 1.4 | 1.3 |
| 1983 | 0.9 | 0.9 | 1.3 | 1.2 | 0.9 | 0.9 |
| 1984 | 1.1 | 1.0 | 1.1 | 0.7 | 0.9 | 0.7 |
| 1985 | 0.7 | 0.6 | 0.7 | 0.6 | 0.7 | 0.5 |
| 1986 | 0.8 | 1.1 | 1.2 | 0.3 | 0.8 | 0.7 |
| 1987 | 1.1 | 0.7 | 0.7 | 0.4 | 0.5 | 0.4 |
| 1988 | 2.2 | 1.1 | 1.3 | 0.6 | 0.8 | 0.8 |
| Total | 1.0 | 1.0 | 1.0 | 1.0 | 1.0 | 1.0 |

# Chapter Five

# International Business Strategies of Nippon Enterprise Development

### Hiroaki Ueda

The Japanese venture capital industry is currently enjoying its third boom cycle due to the changes in the industrial structure, the loosened criteria for OTC stock registration and monetary relaxation. The total equity investment balance of venture capital firms reached 476.6 billion (equivalent to $32 billion) at the end of March 1990, growing by 32% from the previous year. The number of venture capital-backed IPO companies has also increased from 101 in 1988, to 123 in 1989. (See Table 5.1.) This recent boom has enticed many newcomers to enter the venture capital industry. The number of Japanese venture capital firms now exceeds 100. They are aggressively competing with each other to obtain investment opportunities in promising IPO

## Table 5.1
## Japanese VCs—Investment Balance and Number of IPO Companies

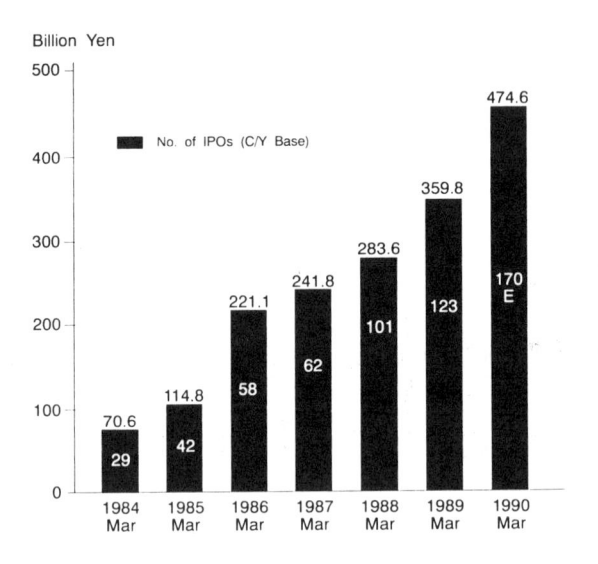

| Year (Mar) | Investment Balance (Billion Yen) | No. of IPOs (C/Y Base) |
|---|---|---|
| 1984 | 70.6 | 29 |
| 1985 | 114.8 | 42 |
| 1986 | 221.1 | 58 |
| 1987 | 241.8 | 62 |
| 1988 | 283.6 | 101 |
| 1989 | 359.8 | 123 |
| 1990 | 474.6 | 170 E |

## Table 5.2
## Japanese VC Firms Classification by Capital Relationship (as of August 21, 1990)

| Affiliate | Number |
|---|---|
| Bank (NED, Tokyo Venture) | 42 |
| Securities Corp. (JAFCO, NIF) | 30 |
| Independent (Techno-Venture) | 10 |
| Foreign Corp. (Schroder PTV) | 3 |
| Trading/Lease/Maker Related | 8 |
| Bank/Securities JB | 3 |
| Others (Gov. Related) | 9 |
| Total | 105 |

candidates, (see Table 5.2) and are now aggressively entering foreign markets.

The history of internationalization at Nippon Enterprise Development Corporation may be quite informative as it relates to the expansion of the Japanese VC industry.

Unlike most Japanese venture capital firms, NED is a multipurpose finance company that provides debt as well as equity investment to the small- and medium-sized companies. NEDs loan business generates a constant and stable income in order to support its equity investment activity.

From 1972 to 1974, which was the first venture capital boom in Japan, eight private venture capital firms, including NED, were established in response to the following factors:

1. The successful business results of U.S. venture capital firms.

2. The contribution of U.S. venture capital to encouraging growth in high technology industries.

NED has been gradually investing in U.S. venture businesses since 1975, mainly through introductions from our shareholders for the following reasons:

1. In order to acquire seed and early-stage investment know-how.

2. NED has been historically entrepreneurial and globally oriented.

In addition to small direct investments, we began participating in venture capital funds and establishing business tie-ups with leading foreign venture capital firms in order to strengthen deal flow and to expand our international network. In August 1983, we invested in the first fund organized by Columbine Ventures based in Denver. In May 1985, a tie-up with GIMV (an investment cooperation of Flanders, Belgium) was signed. We subsequently joined in funds established by two leading U.S. venture capital firms, Kliener Perkins Caufield & Byers and TA associates. (See Table 5.3.) Thus, we attempted to diversify and moderate our foreign investment risks through this combination of direct and indirect investment through local funds.

However, even so, our early foreign investments were far from successful. This is due, in part, to the lack of in-house technical expertise to evaluate potential investment candidates—most of the Japanese venture capitalists are from banks or securities companies and have either accounting or taxation backgrounds rather than engineering or science training. Like all foreign venture capital firms, we were "outsiders" in the U.S. venture industry, making it difficult to access the foreign VC networks. There are significant differences in statutes, business ethics and commercial customs between the U.S. and Japan. For example, in Japan, The Anti-Trust Act prohibits venture capital firms from appointing directors to the boards of investee companies. Thus, we had little experience with hands-on management and monitoring of investments. Finally, NED did not have an internal infrastructure to support international investment activities in the early 1980s.

Until recently, NED relied almost entirely on our shareholders and tie-up partners for deal flow and due diligence. This passive approach to direct investing contributed to the mixed performance of our early investments. Considering these factors, we concluded that localizing our activities by establishing foreign subsidiaries was the appropriate next step. This was intended to enable us to become integrated into the local VC networks and to conduct "primary" due

## Table 5.3 NED Group

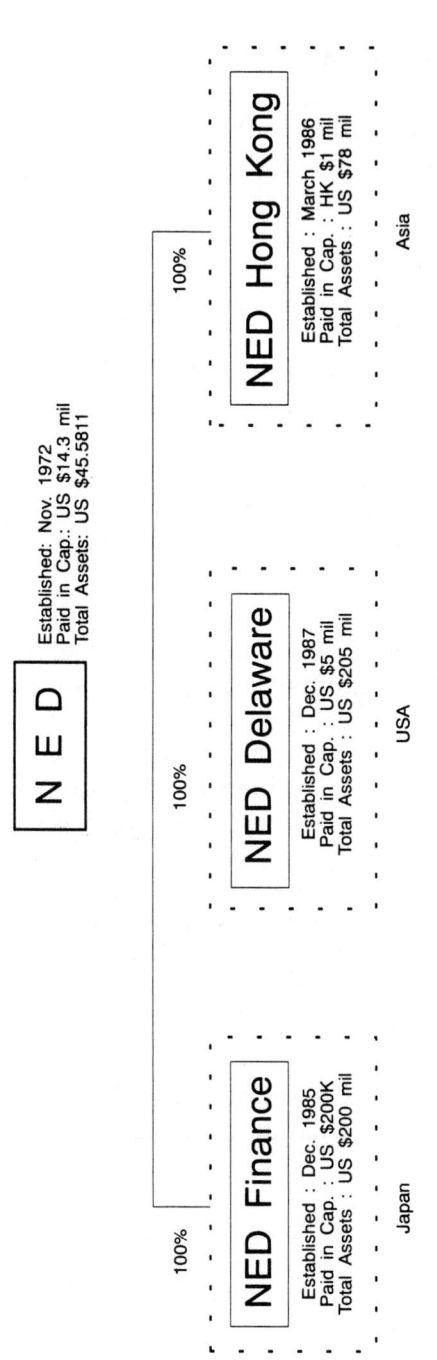

```
                          ┌─────────┐
                          │  N E D  │  Established: Nov. 1972
                          └─────────┘  Paid in Cap.: US $14.3 mil
                               │       Total Assets: US $45.5811
         ┌─────────────────────┼─────────────────────┐
        100%                  100%                  100%
         │                     │                     │
┌─────────────────┐  ┌─────────────────┐  ┌─────────────────┐
│   NED Finance   │  │  NED Delaware   │  │ NED Hong Kong   │
│                 │  │                 │  │                 │
│Established: Dec.│  │Established: Dec.│  │Established: Mar.│
│ 1985            │  │ 1987            │  │ 1986            │
│Paid in Cap.:    │  │Paid in Cap.:    │  │Paid in Cap.:    │
│ US $200K        │  │ US $5 mil       │  │ HK $1 mil       │
│Total Assets:    │  │Total Assets:    │  │Total Assets:    │
│ US $200 mil     │  │ US $205 mil     │  │ US $78 mil      │
└─────────────────┘  └─────────────────┘  └─────────────────┘
      Japan                 USA                  Asia
```

diligence. However, up until recently, these subsidiaries have functioned as merely "booking arms" (i.e., paper companies).

In October 1986, our first foreign subsidiary, NED Hong Kong Co., Limited, was established in Hong Kong to identify venture investment opportunities in ASEAN and NIES countries. Then in 1987, we set up NED Delaware Co., which was founded as a strategic base for direct investing in the U.S.

The U.S. represents the largest, most sophisticated and mature venture capital in the world, so we initially decided to focus our venture investment efforts in the U.S. In September 1989, NED began staffing up its U.S. subsidiary. In order to support the increasing level of foreign equity investment activity and deal with the increasing complexity of foreign loan transactions, NED created two separate divisions within the existing International Group—one for equity and one for debt. This change in our organizational infrastructure will allow us to aggressively expand our presence in the U.S., and soon in Asia.

To enable our U.S. staff to penetrate the local market so that they would have access to the best deal flow and information, we decided to strengthen our ties with several top VC firms. In particular, we succeeded in establishing a joint working relationship with New Enterprise Associates (NEA), in October of 1989, surrounding the sponsorship of their affiliated seed fund, Onset Enterprise Associates (OEA).

For us, NEA was the perfect partner because we shared a common vision and commitment to the internationalization of venture capital. In addition, NEA, like ourselves, is one of the largest domestic VC firms, and has a national network presence. The two unique features of our tie-up with NEA are:

1. NEA allowed us to share office space in their Menlo Park office to foster the sharing of information and opinions between our organizations

2. Our U.S. representatives work closely with the partners of OEA to learn the process of seed investing.

Furthermore, to expand and strengthen our information network in the U.S., we became members of two leading VC associations, the Western Association of Venture Capitalists, and the National Venture Capital Association. Joining these organizations demonstrates our long-term commitment to integrating NED into U.S. venture capital companies and to promoting internationalization.

NED was founded as a multipurpose financing company in Japan. Thus, we have a charter to provide both equity and debt financing to companies in all stages. (See Table 5.4.) In fact, NED is the largest Japanese VC company as measured by the amount of total assets, which currently exceeds U.S. $6 billion.

NED intends to continue as a supplier of both debt and equity financing to venture stage companies in the U.S. market. The ability and commitment to provide debt financing differentiates NED from the other Japanese VCs that are active in the U.S. and even from domestic VCs. Here again, we find it imperative to find a partner to help us identify and evaluate lending opportunities. On the West Coast there are several local banks engaged in lending to venture companies located in the Silicon Valley. We are currently in the process of structuring a co-lending program with one of the best high-tech lending banks. We think joint working relationships with NEA and a leading local high-tech bank provides a solid foundation for our business in the U.S.—although expansion of our presence to the East Coast may be required in the near future.

These partnerships, in the true sense of the word, give NED an "insiders" view of the U.S. venture capital business. A view that we believe no other Japanese VC has yet enjoyed.

It is quite natural for a venture capital company situated in the Far East to have a strategy to increase its investment and loan financing into the rapidly growing countries in the ASEAN region and into the Newly Industrializing Economies. Our company made a tie-up with Korea Kuwait Banking Corporation (KKBC), a merchant bank of the Hyundai group, one of the most influential interest groups in Korea, in December of 1987. Besides KKBC, we have participated in several other Korean venture capital funds including those established under the Small and Medium Business Start-up Act, by securities-backed and independent venture capital companies.

## Table 5.4 From the Start-Up Through Post-Public

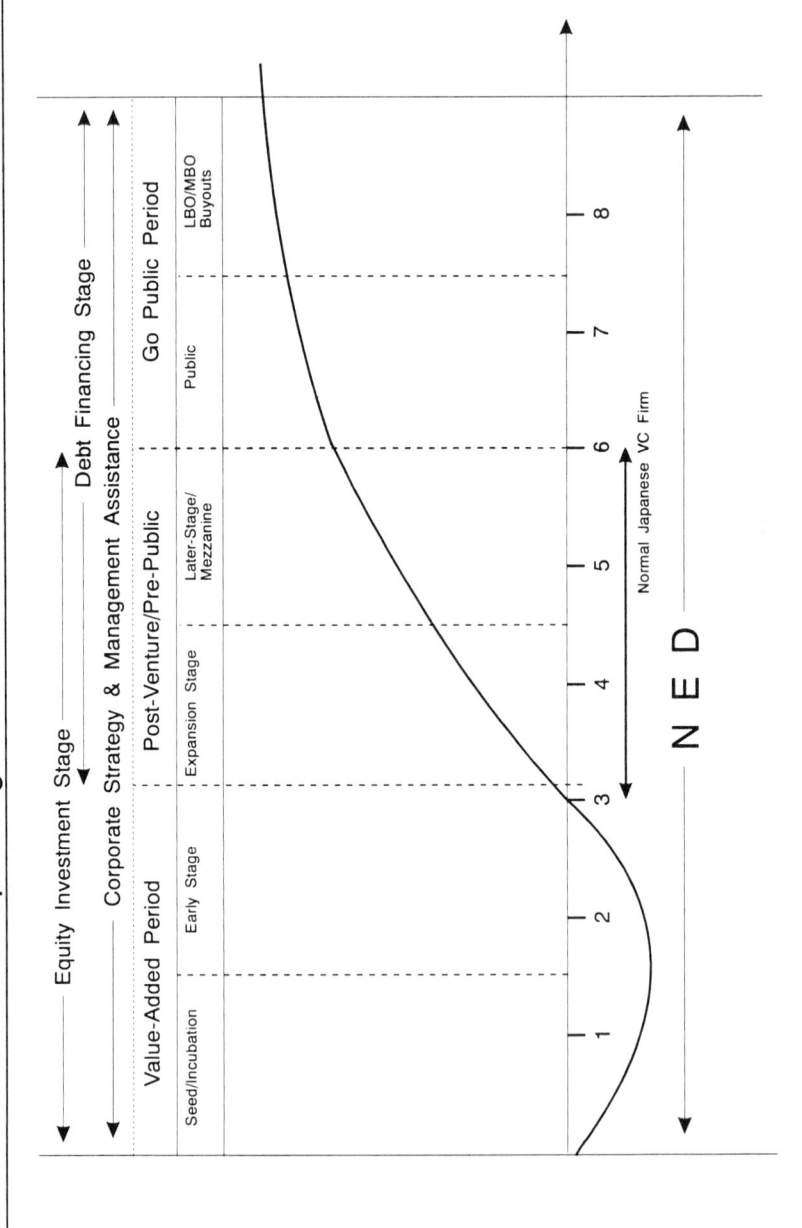

|  | Value-Added Period | | Post-Venture/Pre-Public | | Go Public Period | |
|---|---|---|---|---|---|---|
| | Seed/Incubation | Early Stage | Expansion Stage | Later-Stage/Mezzanine | Public | LBO/MBO Buyouts |
| | 1 | 2 | 3 | 4 | 5 | 6 | 7 | 8 |

Equity Investment Stage → ← Debt Financing Stage

Corporate Strategy & Management Assistance

Normal Japanese VC Firm: 3–6

N E D

In addition, we have an eye on the Hong Kong and Thailand markets. In these countries, stock markets are becoming mature enough for institutional investors, and foreign demand for equities in those local stock markets has expanded due to rapid growth. To date, we have been fortunate to realize capital gains on most of those investments. One trend worthy of note in these countries is the "privatization" of social infrastructure projects (e.g., transportation and distribution systems) through the so-called "Build, Operate and Transfer" (BOT) scheme. These projects represent excellent investment opportunities because they are often financed with foreign capital, yet managed and operated by locals with the goal of "going public" upon completion, usually within three to five years. We have participated in several such projects, particularly in Hong Kong.

We are also eventually planning to expand our activities into such countries as Indonesia and Malaysia. There is no doubt that these countries will have a better environment for public matters in the near future, which enables VC investors to "exit" their investments. We firmly believe that in these countries, export-oriented businesses, tourism and infrastructure projects will create good investment opportunities for foreign venture capitalists. For now we will carry on our efforts to find attractive investment opportunities by making use of our business connections with major local banks, securities companies and investment companies. As a near term milestone for our company's strategy for local market penetration in the Asian countries, we are planning to open a representative office in Singapore.

Companies all over the world must look to foreign markets during the earliest stages of their growth and development. Venture capital firms who are poised to assist these companies will capture superior returns that result from creating value across the borders. NED is attempting to differentiate itself both domestically and in the U.S. as a strategic liaison and financial facilitator of the international expansion of small businesses.

In other words, today's foreign venture capitalists must perform such value-added roles as consultant and/or agent relating to business tie-ups and the establishment of joint-ventures for portfolio companies.

It is too early for us to make predictions about the ultimate outcome of our internationalization efforts. However, we would like to confirm that, as a pioneer of Japanese venture capital firms, we intend to continue all our efforts to be an industry leader with a global vision for today's age.

*Mr. Hiroaki Ueda is President of Nippon Enterprise Development Corporation.*

# Chapter Six

# Japan as a Source of Capital and Debt Financing for U.S. High-Tech Companies

## Robert Brown

There are a number of recent developments that are likely to affect the ability of Japan to act as a source of money for U.S. high-tech companies. On the equity side, as of December 7, 1990, the Tokyo Stock Exchange is down 43.8% this year. By way of comparison, the Hong Kong Stock Exchange, is up 6.1% for the same period.[1] Similar downward spirals can be seen in the Taiwan stock market that had lost 78% of its value in 1990, Manilla 63%, Bangkok 42%, Seoul 39%, Singapore 31% and Kuala Lumpur 21%.[2] On the debt side, it is reported that Japanese banks are encountering difficulties meeting the international capital guidelines.

Despite these rather dramatic changes, many U.S. companies continue to look to Japan as a source of equity and debt financing. Apple Computers, for instance, recently became listed on the Tokyo Stock Exchange.

The advantages for U.S. companies of raising money in Japan go beyond the mere advantage of raising the money itself. Such financing can also help to reinforce important alliances with key financial institutions, expand customer bases, build the confidences of employees, officers, and directors, and even help in recruiting.[3]

Thus, there are many advantages to continuing to use Japan as a source of money. However, there are two factors that may affect this ability—recent economic developments and the extent of securities regulation in Japan.

## Recent Economic Developments

In the past year, there have been a number of significant developments in the Japanese economy that may affect the ability of foreign companies to raise debt or equity from Japanese sources. The one that has received the most publicity is the decline in value of the Tokyo Stock Exchange. At its lowest level, the Nikkei Index of the Tokyo Stock Exchange had fallen 49% since January 1, 1990.[4] As noted, recently it is 43.8% below its value at the beginning of the year.

Looking at land prices, according to the Japanese Construction Ministry, in the August through September 1990 period, second-hand condominium prices in Tokyo, Osaka, and Nagoya fell between 5 and 30% from the preceding three months.[5] Land prices in general have stopped their rapid rise and are reported to be holding steady, if not falling.[6]

For banks, these developments in the stock and real estate markets have had a material impact. Under International settlements, the capital of banks must equal or exceed 7.25% of their loans by April 1, 1991 and 8% by April 1, 1993.[7] Since banks are allowed under these rules to count as capital 45% of the unrealized gains on securities they hold (called hidden reserves), the decline in the Tokyo Stock Exchange has meant a decline in the capital of Japanese banks.[8] A decline in capital, in turn, means a reduction in lending capacity in order to comply with the guidelines.

According to one report, every drop of 1,000 points on the Tokyo Stock Exchange increases the aggregate city bank capital shortage by about ¥.5 trillion and reduces their capital adequacy ratio by .2%.[9] It is reported that the ratio for The Dai-Ichi Kangyo Bank fell from 8.3% on March 31 of this year to 7.4% on September 30, 1990.[10] According to another report, its ratio may be as low as 6.8%.[11] For the same period, The Fuji Bank will probably report a 7.5% ratio down from 8.2%, while The Industrial Bank of Japan fell from 7.8% to 7.5%.[12] These ratios compare unfavorably with those of many Western banks. Deutsche Bank had a ratio in mid-1990 of 11.5% while Britain's Barclays PLC was at 9.0%. J.P. Morgan and Company was around 10.3%.[13] Due to these problems, there is some discussion about the capital adequacy guideline ratios being adjusted in light of world economic conditions.

Making matters worse, the decline in capital as a result of the decline of the value of the hidden reserves will highlight the traditionally low profitability of Japanese banks. The so-called Japanese city banks produced an average return on assets of .24% in the fiscal year ending March 1990 while U.S. money-centered banks generated an average return on assets of .98%.[14]

As a result of their difficulty in meeting the guidelines and their low profitability, Japanese banks have begun to fall into a vicious cycle. It has been increasingly more difficult for them to raise money in the international markets to meet the capital adequacy guidelines. Specifically, it has been reported that they are paying an eighth to a quarter of a percentage point more than their competitors in the Euro-dollar interbank market and the markets for Euro-dollar certificates of deposits and commercial paper. Triple A ranked European banks such as Germany's Deutsche Bank pay 25% basis points less than that paid by the Japanese.[15]

There is even some doubt as to the ability of the regional banks, thrifts and small cooperatives, known as "shinkin" in Japan, to avoid bankruptcy. An official at The Bank of Japan in charge of supervising small financial institutions was reported in October of this year as saying that although no financial institutions are in immediate danger of going bankrupt, there is a chance that some will in the future.[16]

Further compounding the problems facing Japanese banks is the rapid rise of interest rates. In May of 1989, the official discount rate of The Bank of Japan stood at 2.5%. By August 30 of this year, it had risen to 6%.[17] This has made it difficult for Japanese banks to attract new borrowers so as to continue their rapid growth. In light of their capital problems, this may be a blessing since one way of meeting the guidelines is by reducing loans.

Another problem facing Japanese banks is the crackdown by the Ministry of Finance on real estate related loans. In October 1989 and again in March of 1990, the Ministry of Finance introduced a series of monitoring measures designed to decrease real restate loans.[18] The restrictions called on banks to refrain from lending for speculative land transactions and to closely investigate the purpose of their loans. Banks were also required to restrain lending to real estate firms to a level lower than their then existing total lendings to such firms and to disclose to the Ministry the amount of their loans to real estate, construction and non-banking financial firms. Again, however, for the reasons noted, this reduction in loans may be a blessing.

The securities companies of Japan have similarly been beset with problems reflecting the condition of the Japanese economy. They have encountered tremendous losses as a result of the general decline of the Japanese stock market. All of the Big Four securities firms in Japan reported sharply diminished sales and profits for the period ending September 30, of this year.[19] According to *The Wall Street Journal*, as a result, the capital pools of some of Japan's biggest securities companies shrank in 1990 between 20 and 30%.[20] The Finance Ministry changed its rules concerning the appraisal of losses for stock held in securities investment accounts so that such losses could be treated as extraordinary losses and thus excluded from pretax profits.[21] Otherwise, it was reported that almost all of the major securities firms would have reported losses for that period instead of mere reductions in profits.

On the corporate side, although little recognized outside of Japan, companies during the 1980s shifted to a greater dependence on profits and new capital issues rather than on bank loans as a source of funds. In a sense, they had distanced themselves from banks as sources of funds. If the pattern had continued, they could have been isolated from the problems facing Japanese banks. But this pattern has recently changed. In 1990, the reserve of liquid assets of Japanese companies began to fall for the first time in many years. This will mean that they may have to return to the banks as providers of funds. If so, they will have to pay more. Banks may, therefore, be less willing to lend to foreign high-tech companies as major Japanese companies return to them as borrowers.

## Securities Regulations and Insider Trading

Another problem facing individuals wishing to raise capital in Japan is the degree of securities regulation and insider trading in Japan. Effective August 23, 1988, Article 154 of the Japanese Securities and Exchange Law was revised so as to allow the Finance Ministry to ask the Tokyo Stock Exchange to check the records of listed companies involved in suspicious transactions.[22] From April 1, 1989, insider trading became punishable as a criminal rather than a civil offense.

New regulations that became effective on April 1, 1989, prohibit investors with access to material company information from trading shares, including convertible bonds and bonds with warrants, before the information was made public. Those affected fall under two categories of insiders and quasi-insiders. The first group includes company executives, employees and major stockholders with holdings of 10% or more. Quasi-insiders are government officials, others having legal authority over companies, outside contractors such as lawyers, certified public accountants, underwriting securities companies, main banks and journalists. Material information includes information on financing new share issues, changes in dividend payments, merger plans, new products, losses caused by accidents, and changes in major stockholders and business forecasts. However, information about business forecasts would only be material if it is likely to change a company's sales by more than 10% and recurring after-tax profit by more than 30%.

New guidelines prohibit company insiders from trading shares until 12 hours after material information has been made known to two or more new agencies. Violations of these restrictions under Article 200, paragraph 4 of the revised Securities and Exchange Law would carry a fine of ¥500,000 or six months in jail.

Notwithstanding this beefing up of insider regulation, the general opinion of many foreigners in Japan is that insider trading is viewed with as much alarm in Japan as "illegal parking or jay-walking."[23] Others have called Japan's attitude towards enforcement of insider trading laws as examples of "benign neglect."[24] As the Director General of the Securities Bureau of the Ministry of Finance stated earlier this year, the Ministry is not "set-up to expose insider trading."[25]

Not until April of this year did the Tokyo Metropolitan Police begin its first in-depth investigation of insider trading by seizing records in connection with the Nisshin Kisen stock manipulation case.[26] Prior to that time, no one had been formally investigated or prosecuted in Japan for insider trading. In the United States by contrast, in 1989, 42 insider cases were prosecuted and more than $60 million in fines were paid. Another 1,000 cases were under investigation by about 2,000 members of the U.S. Securities and Exchange Commission.[27] Until recently, Japan had 17 Finance Ministry officials assigned to investigate insider trading with none of them having any real investigative or enforcement powers.[28] While insider trading can be punished in the United States with a maximum term of five to 10 years and fines ranging from $500,000 to $2.5 million, the maximum punishment in Japan, as noted, is only six months in jail and a maximum fine of ¥500,000 or approximately $3,500.[29]

Some Americans in the field have alleged that insider trading in Japan is a non-tariff barrier. According to this argument, Japanese securities firms are more inclined to participate in such insider trading, whereas their American counterparts are not. This allegedly gives Japanese firms an unfair advantage.[30] One individual has even reported that the financial laws are selectively enforced against firms that are not part of the inner circle of Japanese firms and that this selective enforcement constitutes a barrier to U.S. companies doing business in Japan.

## Future Developments

As a result of recent economic developments in the U.S. and Japan and the present state or condition of securities regulations in Japan, it is possible that we may see a number of important developments over the next year as Japanese banks and securities companies respond to these conditions.

As a first step, in order to comply with the capital adequacy regulations of The Bank for International Settlements, unless those guidelines are revised, major banks in Japan will use subordinated loans to raise capital.[31] By issuing such subordinated loans, the banks would raise what is called Tier 2 Capital. The primary borrowers of such loans would be city trust and long-term credit banks while the main lenders would be insurance companies. In essence, the insurance companies would be lending money to the banks in order to enable the banks to meet the capital adequacy guidelines. To a certain extent, this has already begun to occur. Sumitomo Bank has been allowed to issue subordinated bonds in the public markets in the United States and Europe in order to meet the guidelines. It was reported that since July of 1990, nearly a dozen banks either have planned or completed large issues of subordinated loans.[32]

Another action being taken by banks to comply with the capital adequacy guideline ratios is selling assets. According to one source, the 12 largest commercial banks in Japan will have to slash their assets by 16% to repair the damage caused by the decline in the Tokyo stock market unless they are able to raise $36.5 billion in new capital.[33] According to the *International Financing Review*, several senior officials of Japan's commercial banks have stated that they have started to liquidate assets at home and abroad to sharply curtail balance sheet losses.[34] Such bankers have also been reported as saying that they plan to trim back interbank funding with banks overseas as well as holdings of foreign bonds. It has been reported that Japanese investors have reduced their holding of U.S. stocks and bonds by almost $9 billion in the first six months of this year.[35] Some lending specialists have indicated that the assets held by Japanese banks in the form of commercial loans and mortgages are selling at discounts of as much as 10% on some syndicated loans.[36] At a minimum the growth of

Japanese banks' assets will continue to ease. For instance, in the first quarter of this year according to The Bank for International Settlements, the growth of Japanese banking assets overseas grew by $19.4 billion from the fourth quarter of 1989 but that this was down from a gain of $25.2 billion over the preceding quarter.[37]

Similarly, the growth of the Japanese securities companies has also been reduced. They have traditionally been dependent on three basic sources of income: stock and bond trading commissions, stock and bond underwriting commissions, and in-house trading of securities. As the Tokyo Stock Exchange volatility has increased this year, trading volume has fallen by nearly half from a year ago. This means that commission income is down sharply. Similarly, underwriting fees were reduced in part earlier this year because of a self-imposed ban by the securities industry on new issues of stocks and equity-linked bonds. Even in-house trading revenues will fall sharply because of the evaporation of warrants—one of the hottest profit centers for Japanese securities companies.[38]

## Effect on U.S. Companies

The effect on the United States in general and high-tech companies in particular of these developments and conditions may be far reaching. First, gone are the days of cheap yen. Japanese banks are charging more as a result of the rise in the discount rate and their increasing costs in raising capital in the U.S. and Europe. Similarly, the downgrading of the credit rating of Japanese banks as a result of these events has raised their cost, for instance, for the issuance of letters of credit to secure municipal bonds. As a result of the recent downgradings of the credit rating of The Fuji Bank and The Dai-ichi Kangyo Bank from AAA to AA-1 by Moody's Investors Service, interest rates for such letters will be raised for such banks by four or five basis points. This will add $50,000 interest a year to a $100 million bond.[39]

Second, foreign borrowers will in essence be facing restricted lending ability by Japanese banks as the banks strive to meet the capital adequacy guidelines. To do so they will be limiting loans, selling assets and raising money on their own from Japanese insurance companies and in Europe and the U.S.

Third, foreign borrowers will be faced with restrictions by the Ministry of Finance on loans for real estate purposes, and will be competing with domestic corporate borrowers who will again be borrowing money from Japanese banks instead of internally generating it or raising it themselves.

## Putting these Developments in Perspective

In conclusion, the ability of Japan to provide cheap capital and debt for the world, including high-tech companies, is decreasing. But this change must all be put in perspective. Six of the largest banks in the world are still Japanese.[40] And the Japanese economy is expected to grow at a rate of around 5% or 6% this year, well above that of many other countries in Europe and Asia. Even the efforts in Japan to raise interest rates point out that while the rest of the world is concerned about recession, Japan is primarily concerned with inflation.

It is likely then that instead of continuing their tremendous growth, Japanese banks and securities companies will hold or maintain their present position in the world's international money markets. Thus, Japan will remain an important source of capital for emerging companies. The difference is that Japan will be competing with the same relative strengths as their U.S. and European competitors. They will not be necessarily more competitive.

*Robert Brown is a Resident Partner of the Tokyo Office for Pillsbury Madison & Sutro.*

# Endnotes

1. "Tokyo Stocks Finish Slightly Below Day's High; London Shares Gain on Hope for Gulf Settlement," *The Wall Street Journal* (December 7, 1990), World Markets, Pg. C10.

2. "Tokyo Exchange's Mood Is Gloomy," *San Francisco Examiner* (November 11, 1990), Business, Pg. 1.

3. "Going Public, Japanese Style," *The Wall Street Journal* (May 2, 1988).

4. "Japan May Delay Deregulation 10 Years," *American Banker* (November 7, 1990), International News, Pg. 18.

5. "Land Prices Ease in Buyers' Market," *The Japan Economic Journal* (November 17, 1990), Pg. 1.

6. *Ibid.*

7. "Japanese Banks Sneeze, World Catches Cold?", *Japan Economic Institute Report*, (Washington D.C.: Japan Economic Institute, November 16, 1990), No. 44B, Pg. 6.

8. "Japan's Banks, Too, Want Right to Securitize," *American Banker* (November 13, 1990), Finance, Pg. 24.

9. "Japanese Banks Hit by Tokyo Market's Dip," *American Banker* (September 5, 1990), Finance, Pg. 32.

10. "Hard Times Separate Japan's Strong, Weak," *American Banker* (October 29, 1990), Finance, Pg. 16.

11. "Woes Cutting Bank Capital as Tougher Rules Near," *American Banker* (October 4, 1990), Finance, Pg. 1.

12. "Hard Times Separate Japan's Strong, Weak," *American Banker* (October 29, 1990), Finance, Pg. 16.

## Japan as a Source of Capital and Debt Financing 81

13. "Woes Cutting Bank Capital as Tougher Rules Near," *American Banker* (October 4, 1990), Finance, Pg. 1.

14. "Hard Times Separate Japan's Strong, Weak," *American Banker* (October 29, 1990), Finance, Pg. 16.

15. "Funds Costlier for Japan's Banks," *American Banker* (October 16, 1990), Pg. 1.

16. "Japanese Bank's Troubles Bode Ill for United States," *American Banker* (October 3, 1990), Comment, Pg. 4.

17. "Japan's Banks on Shaky Ground," *San Francisco Examiner* (October 14, 1990), Business, Pg. E1.

18. "MOF Seeks to Further Tighten Land Loan Rules," *The Japan Economic Journal* (November 17, 1990), Pg. 5.

19. "Japanese Securities Firms Record Big Profit Declines Over Six Months," *The Asian Wall Street Journal Weekly* (October 29, 1990), Pg. 24.

20. "Capital Dries Up Japan's Securities Firms," *The Wall Street Journal* (October 23, 1990), Pg. C1.

21. "Japan Alters Rules to Cushion Earnings of Securities Firms," *The Asian Wall Street Journal Weekly* (October 22, 1990), Pg. 29.

22. "Investigators Tightening Up on Insider Trading," *Kyodo News Service* (August 16, 1988).

23. "Insider Trading Prosecution a First for the Japanese," *Chicago Tribune* (May 13, 1990), Business, Pg. 6.

24. "Authorities to File Japan's First Insider Trading Charges," *The Japan Economic Journal* (May 5, 1990), Pg. 32.

25. "Insider Trading Prosecution a First for the Japanese," *Chicago Tribune* (May 13, 1990), Business, Pg. 6.

**82**  Chapter Six

26. "Authorities to File Japan's First Insider Trading Charges," *The Japan Economic Journal* (May 5, 1990), Pg. 32.

27. *Ibid.*

28. "Insider Trading Prosecution a First for the Japanese," *Chicago Tribune* (May 13, 1990), Business, Pg. 6.

29. *Ibid.*

30. "When It Comes to Fraud, The Gaijin Are the Last to Hear," *Institutional Investor* (June 1990), Japan Journal, Pg. 37.

31. "Banks May Soon Use Subordinated Loans to Raise Capital," *The Asian Wall Street Journal Weekly* (June 11,1990), Japan, Pg. 5.

32. "Chase Bid to Raise Capital in Japan Runs Into Snags," *The Wall Street Journal* (September 17, 1990), Pg. C12.

33. "Huge Capital Deficit Seen In Japan; Stock Slump Leaves the Banks $36.5 Billion Short," *American Banker* (October 2, 1990), International News, Pg. 10.

34. "Global Dispatches: Leading Japanese Banks Seen Liquidating Assets," *American Banker* (August 23,1990), International News, Pg. 14.

35. "Japan's Cash Pool Shrinks, Pinching Credit Globally," *The Asian Wall Street Journal Weekly* (September 24, 1990), Pg. 2.

36. "Bank's Fire Sale of Loans Floods World Markets," *American Banker* (October 3, 1990), Pg. 1.

37. "Stock Fall May Curb Bank Growth Abroad," *The Japan Economic Journal* (September 8, 1990), Pg. 1.

38. "Japan's Securities Firms Await Aftershocks," *The Wall Street Journal* (April 6, 1990), Pg. C1.

39. "Japan Bank Woes Raise Interest Cost for U.S. Towns," *The Japan Economic Journal* (September 8, 1990), Money & Investment, Pg. 30.

40. "Europe's Surging Banks Outspace Japanese, U.S.," *American Banker* (July 26, 1990), Pg. 1.

# Chapter Seven

# The Over-the-Counter Market in Japan

## Hidemi Suzuki

In just the past few years, the over-the-counter (OTC) market in Japan has rapidly expanded in terms of size and function. This market has developed into an active marketplace for raising funds by issuing both equity and debt. And today, going public on the Japanese OTC market rather than on the stock exchanges has become one of the most effective ways of raising large sums of money.

Our main concern today for the purposes of this chapter is the over-the-counter stock market. In view of this, the major topics to explore include: (1) the history and growth of the OTC market, (2) the market today compared to the Japanese stock exchanges as well as to the U.S. OTC market, (3) liquidity and pricing, (4) the issuers and underwriters who participate in the OTC market, and (5) the growing importance of initial public offerings in Japan and their relationship to the OTC market.

I will also examine foreign IPOs and who is participating in this growing trend in Japan's capital markets. Finally, I will briefly outline the current registration procedures for the OTC market.

Overall, the OTC market has historically participated in the rapid growth and development of Japan's securities markets due to its modernizing financial markets and liberalization of its securities laws.

The OTC market in Japan dates back to 1963. At this time, the present country-wide OTC registration system was first introduced to facilitate and organize the post-war boom in equities. It was also during this time that Japan's securities markets started to experience an increase in the activity of non-listed stocks by companies conducting initial public offerings. Since then, the restructuring of Japan's financial markets has favored the growth of the OTC market.

For example, modern trends in corporate finance and changes in the traditional means of raising funds have given Japanese companies a broader range of funding options. These include two fundamental changes in corporate finance: (1) a change to issuing equity publicly at market prices instead of privately at par and (2) a shift from indirect to direct financing.

In 1969, the public subscription of equity issues at market prices began, thus facilitating an increase over the years in initial public offerings in Japan. Today, most of the corporate equity issues are market value issues.

During the 1970s, another important change was the shift to direct financing by issuing equity and debt to raise funds instead of borrowing from banks. This trend, including public offerings of straight and convertible bonds, has taken on a greater importance in the Japanese securities market for two reasons: (1) an increase in the need by Japanese firms for long-term capital and (2) an increase in individual investors participating in the securities markets.

When the Second Section of the TSE was established in 1961, many of the stocks that were originally traded over-the-counter became listed on the Tokyo Stock Exchange. This affected the number of companies trading over-the-counter and later growth of this market. However, in 1976, the Japan Over-the-Counter Securities Co., Ltd., was officially established. The development of this organization was necessary for two reasons: (1) to facilitate and organize the trading of the OTC stock and (2) to ensure fair trading to protect investors.

The liberalization of Japan's securities laws and the modernization of its financial system during the 1980s further contributed to the growth of the OTC market. Various regulatory and technical improvements were developed during this time.

From 1983 to 1985, the government enacted several important regulations to specifically enhance the primary equities market and to allow opportunities for small- and medium-sized companies not listed on the exchanges to raise funds. These included: (1) relaxing restrictions of the dealings and sales of stock in the OTC market, (2) easing the standards of OTC registration and disclosure requirements and (3) revised regulations concerning market making. Furthermore, in order to regulate these standards, a formal organized registered dealer system for OTC stocks was organized in 1983.

The Ministry of Finance (MOF) also continued to contribute to both the development and integrity of the OTC market by encouraging the public offering of new shares and the issuance of convertible bonds. For example, in December 1988, the MOF and JSDA (Japan's Securities Dealers Association) revised the method of determining the offering price of shares and introduced a system of auctioning part of the shares slated for going public. This "public auction pricing system" was enacted to regulate the prices of all newly issued shares or existing shares to ensure fair trading and to protect investors.

Since 1989, the government has also permitted registered shares to be used as security for credit in equity and bond futures transactions.

In addition, the MOF and JSDA proposed this year to enact further measures to ensure fair trading in the OTC market. The proposals to include regulations of insider trading will be directly applicable to the OTC market that up to now have only been enforced on the stock exchanges.

The technical improvements to improve the OTC secondary market included the "QUICK" system, an automatic quotation transmission system developed in July 1984. The QUICK system permitted the use of real time quotations of all OTC issues along with an OTC information system. And in 1989, the three OTC market centers, Tokyo, Osaka and Nagoya, were combined into a single quotation system, thus linking all the major OTC districts and greatly enhancing the efficiency of this market.

Finally, this year, the JSDA introduced the new JASDAQ system (Japan's Securities Dealers Automatic Quotation System), a computerized trading system that will be similar to NASDAQ in the United States.

## Present Role of the OTC Market

Based on the history and growth of the OTC market, the OTC market today functions to serve two roles: (1) to provide financing for smaller corporate issuers and (2) to provide the investing public with a marketplace for trading these issues.

Specifically, the OTC market gives smaller companies the opportunity to issue new stock, raise funds and legally register their stock with the JSDA that regulates the stock market.

## The OTC Market versus the Stock Exchanges

In 1990, the total market value of the OTC market grew to 84% of the TSE Second Section. This is quite an increase compared to when it was 60% in 1989 and only 29% in 1988. (See Figure 7.1.) In looking at Figures 7.1 and 7.2, one can clearly see that the OTC market value has steadily increased between 1984 and 1988. The market value then dramatically increased over 50% in just a two-year period between 1988 and 1990.

Figures 7.3 and 7.4 compare the market value distribution of the TSE with the OTC market as of the end of 1990. Figure 7.3 shows that there is relatively little difference in the market value distribution between the OTC market and the TSE Second Section.

And Figure 7.4 shows this difference in comparison to the significantly larger market value of the TSE First Section.

The Over-the-Counter Market in Japan 89

**Figure 7.1 OTC Market Value as a Percent of the TSE Second SEC**

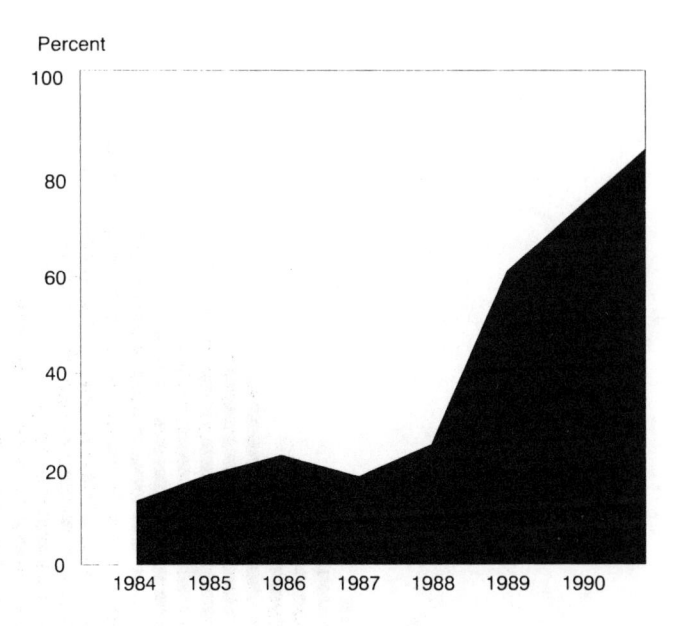

The Nikko Securities Co. Int'l., Ltd.

## 90 Chapter Seven

### Figure 7.2 Total Market Value of OTC Stocks

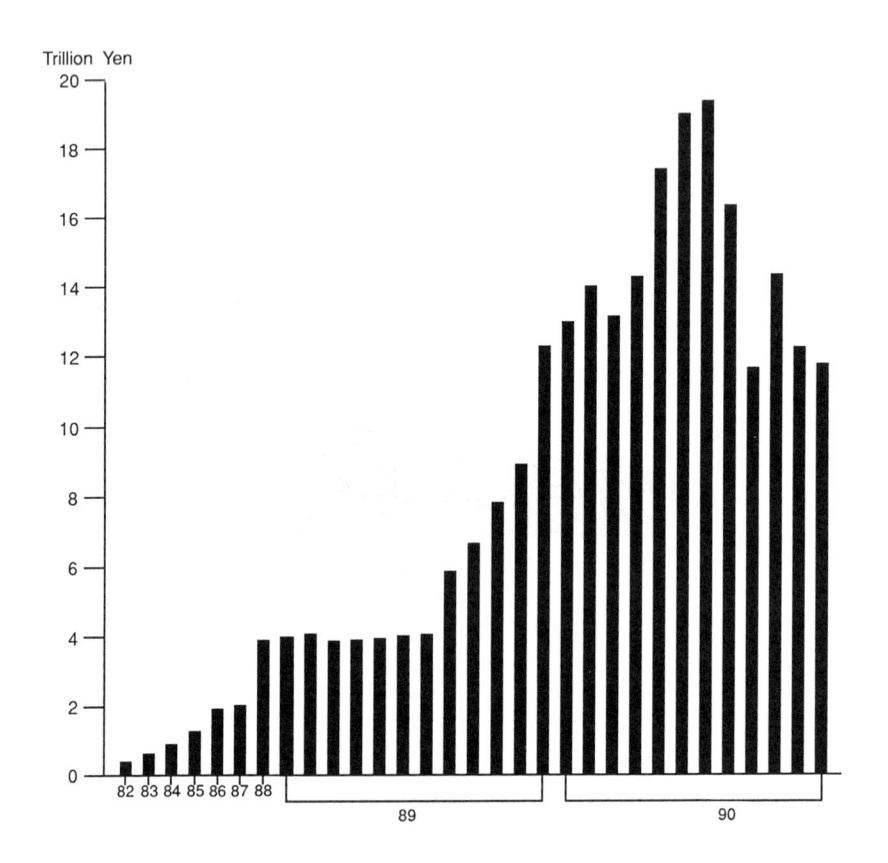

The Nikko Research Center

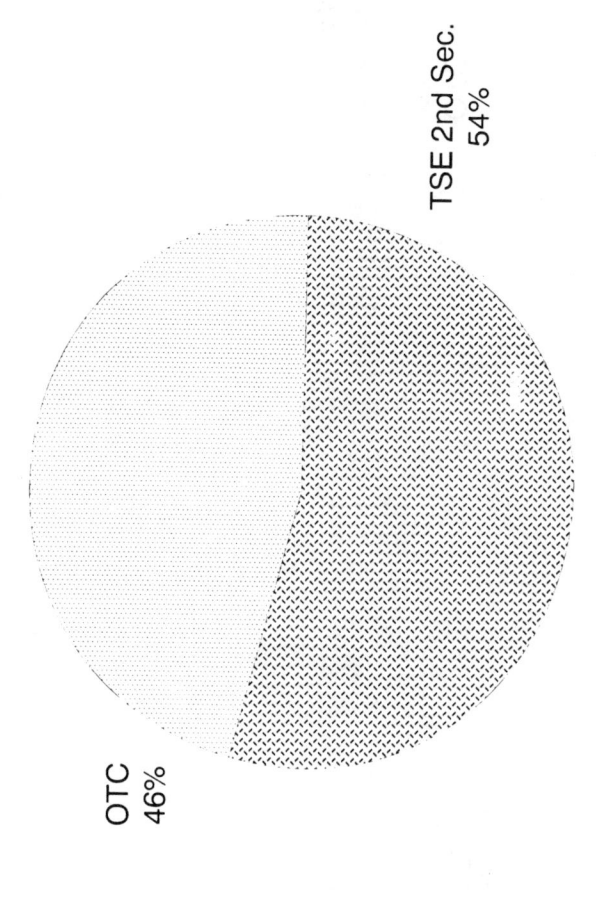

**Figure 7.3 Market Value Distribution TSE Second Section and OTC**

# Figure 7.4 Market Value Distribution TSE and OTC

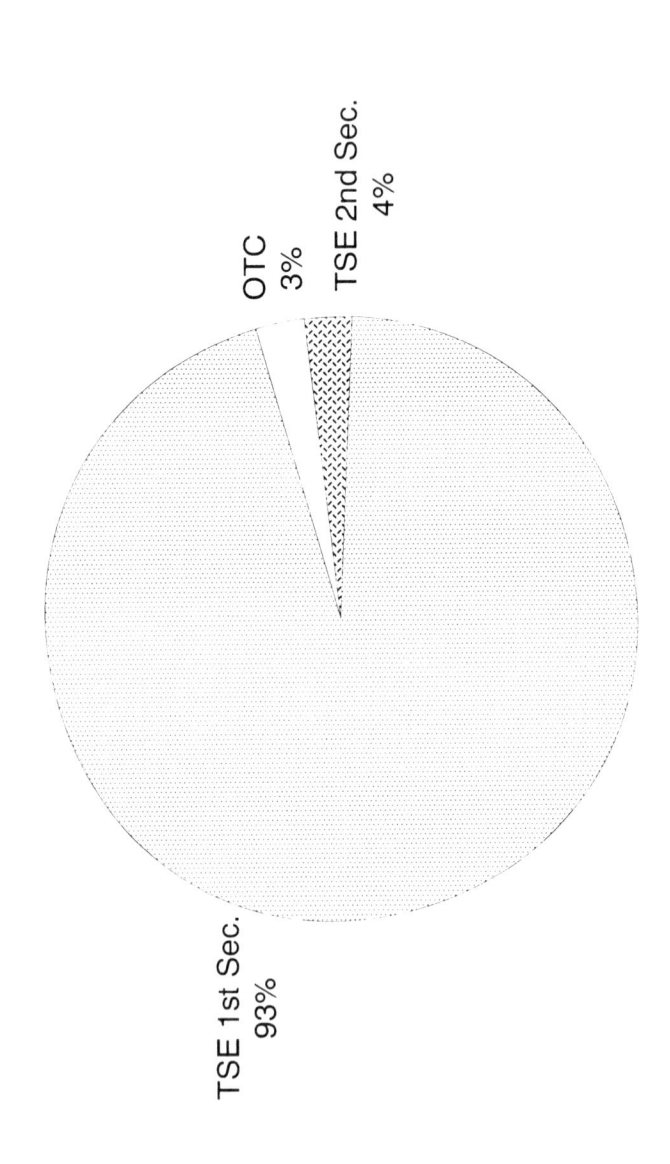

OTC
3%

TSE 2nd Sec.
4%

TSE 1st Sec.
93%

The Nikko Securities Co. Int'l. Inc.

The distinct change in the total market value of the OTC stock coincided with the substantial climb in the OTC stock average, an index similar to the Nikkei 225 Average. Since the end of 1988, when the index bottomed out at 1,238.01, the index rose to 1,355.80 in March 1989. Since March 1989, in addition to the dramatic rise in total market value, the OTC index rose sharply to a peak of 4,149.20 in July 1990.

Some of the reasons why the total market value of the OTC market has sharply risen in just the past two years is due to the sharp rise in the OTC stock index as well as the increase in the total number of registered OTC companies.

In 1990, there were a total of 342 companies registered on the OTC market, a sharp increase from the 263 registered in 1989, and the 196 registered in 1988. (See Figure 7.5.) In comparison, the total number of stocks listed on the TSE Second Section actually decreased slightly between 1988 and 1989, and remained the same between 1989 and 1990. (See Figure 7.5.)

Since 1989, there have been more companies registered on the OTC market than listed on the Second Section of TSE. Figure 7.6 clearly shows that in both 1989 and 1990, there were more new registrations on the OTC market than listings on the TSE Second Section.

In terms of volume shares, the OTC market has also shown a rapid increase in total share turnover in the past few years. In 1990, the total number of shares traded in the OTC market dramatically increased to about 1.1 billion shares, up from 529 million shares traded in 1989. (See Figures 7.7 and 7.8.) Again, in comparison to the TSE Second Section, the total volume of shares more than doubled on the OTC market from 1989 to 1990 while it decreased on this stock exchange. (See Figures 7.7 and 7.8.)

Based on the information shown here, it is apparent that the OTC market has started to receive more attention as a highly recognizable marketplace, especially in the past few years.

One of the reasons why the OTC market is getting more recognition is the criteria used to evaluate the investment quality of the securities registered on this market. Two of these criteria are the Price Earnings Ratio (PER) and the Price Book Ratio (PBR).

## Figure 7.5 Total Companies on the OTC and TSE Second Section

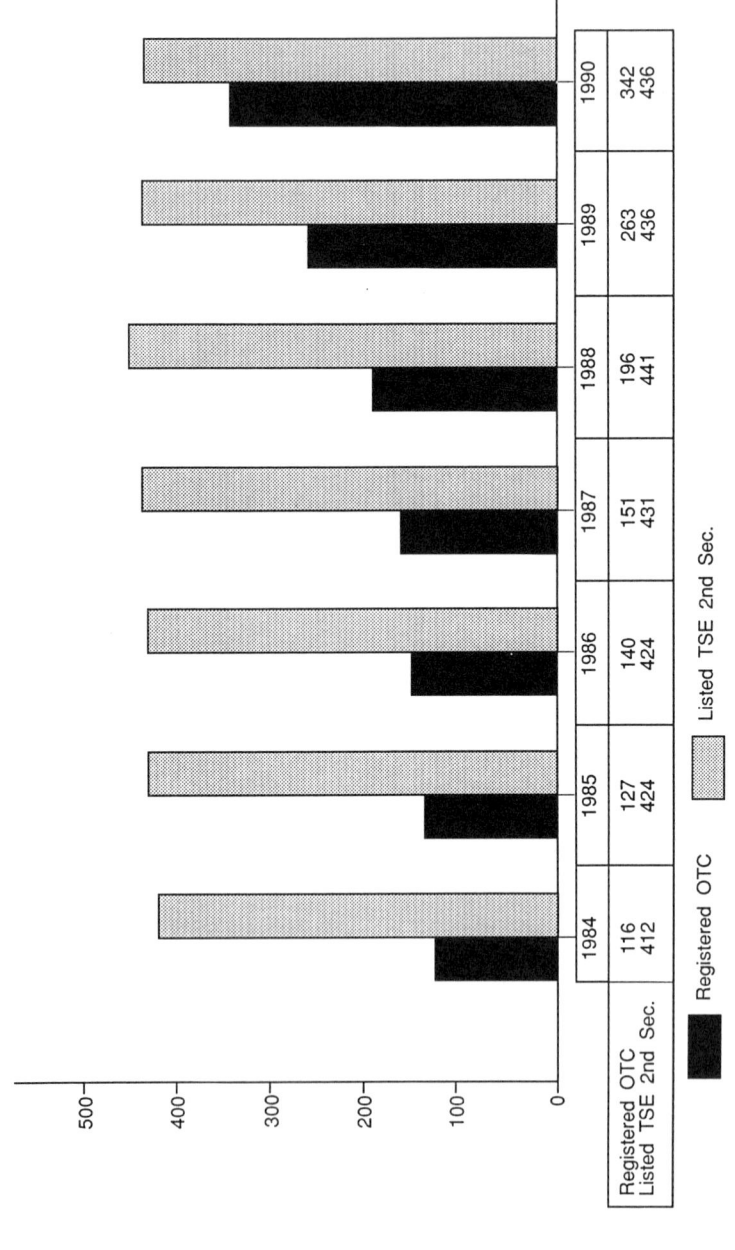

|  | 1984 | 1985 | 1986 | 1987 | 1988 | 1989 | 1990 |
|---|---|---|---|---|---|---|---|
| Registered OTC | 116 | 127 | 140 | 151 | 196 | 263 | 342 |
| Listed TSE 2nd Sec. | 412 | 424 | 424 | 431 | 441 | 436 | 436 |

■ Registered OTC ▨ Listed TSE 2nd Sec.

The Nikko Securities Co. Int'l. Inc.

## Figure 7.6 New Listings/Registration in Japan

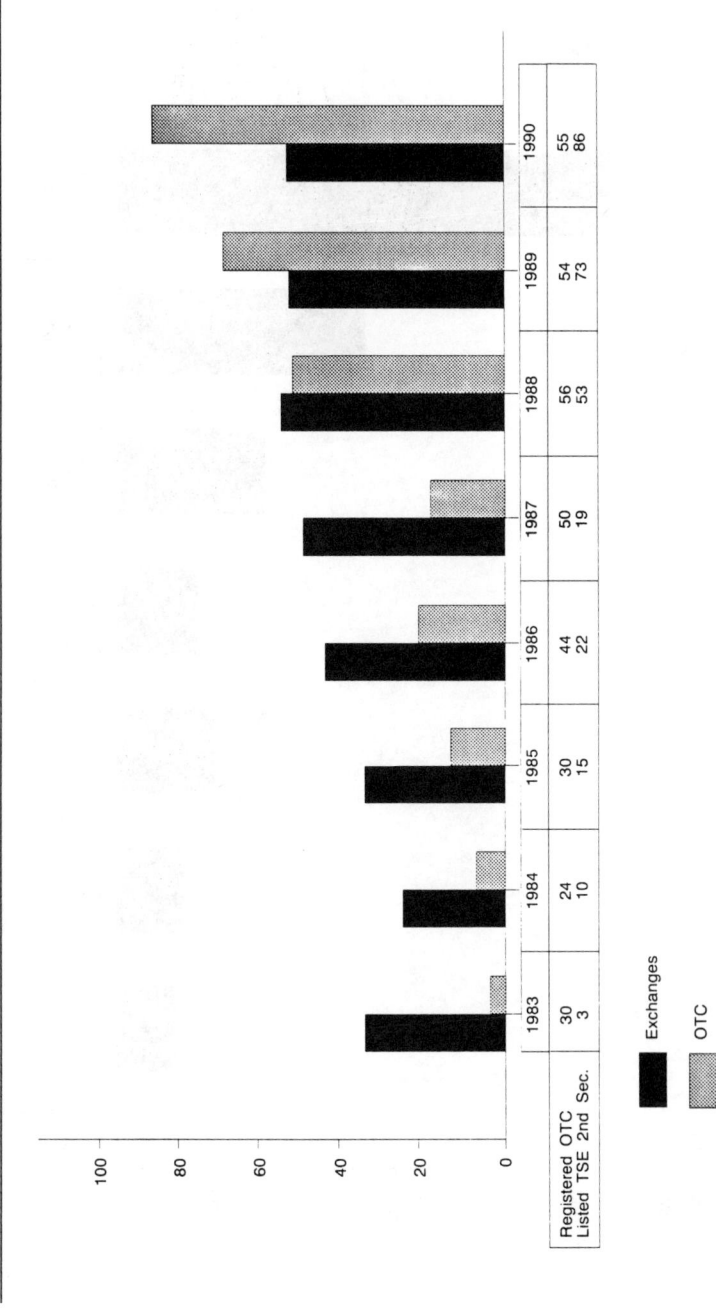

| | 1983 | 1984 | 1985 | 1986 | 1987 | 1988 | 1989 | 1990 |
|---|---|---|---|---|---|---|---|---|
| Registered OTC | 30 | 24 | 30 | 44 | 50 | 56 | 54 | 55 |
| Listed TSE 2nd Sec. | 3 | 10 | 15 | 22 | 19 | 53 | 73 | 86 |

■ Exchanges
▨ OTC

The Nikko Securities Co. Int'l. Inc.

**Figure 7.7 The Japanese OTC Market**

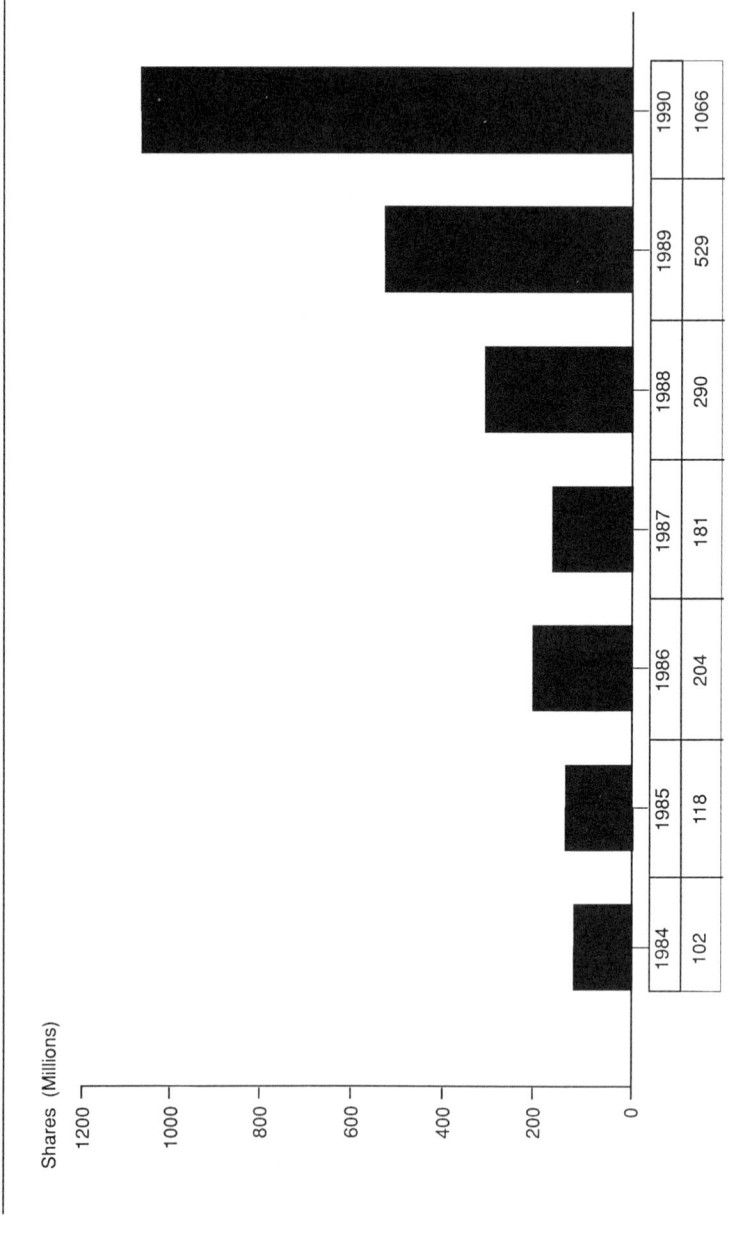

Shares (Millions)

| | 1984 | 1985 | 1986 | 1987 | 1988 | 1989 | 1990 |
|---|---|---|---|---|---|---|---|
| | 102 | 118 | 204 | 181 | 290 | 529 | 1066 |

The Nikko Securities Co. Int'l. Inc.

## Figure 7.8
### Share Turnover (Thousands)

| Year | OTC | TSE 2nd Section |
|------|-----|-----------------|
| 1984 | 102,032 | 4,501,000 |
| 1985 | 118,206 | 3,657,000 |
| 1986 | 203,722 | 4,097,000 |
| 1987 | 181,445 | 4,200,000 |
| 1988 | 290,103 | 4,029,000 |
| 1989 | 529,276 | 4,247,000 |
| 1990 | 1,065,538 | 4,064,000 |

The Nikko Securities Co. Int'l, Inc.

The average price earnings ratio for OTC stocks at the end of 1990 was 59%, which was much higher than the Nikkei Average at 39% and the TSE Second Section Average at 38%. (See Figure 7.9.) In looking at Figure 7.9, which compares the price/earnings ratios (P/E) for Japan's stock market indices, the average P/E ratio for OTC stock has been higher than the stock exchange P/E ratios in the past two years, or specifically at about 75% in 1989 and 59% in 1990.

The Price Book Ratio is another measurement for making investment decisions. The PBR for the OTC market also has been higher than the PBR for the stock exchanges. As of April 11, 1991, the PBR for the OTC market was 7.38 compared to 3.50 for the TSE First Section and 3.27 for the TSE Second Section.

## The OTC Market versus NASDAQ

In addition to comparing Japan's OTC market and stock exchanges, I would like to briefly review the United States over-the-counter market in comparison to Japan's OTC market.

Similar to the U.S. OTC market, small- to medium-sized companies trade on the OTC market in Japan. However, there is a notable difference in the number of registered companies between the two markets; NASDAQ had over 4,100 registered companies in 1990

98   Chapter Seven

## Figure 7.9   Price Earnings Ratios Japanese Stock Indices

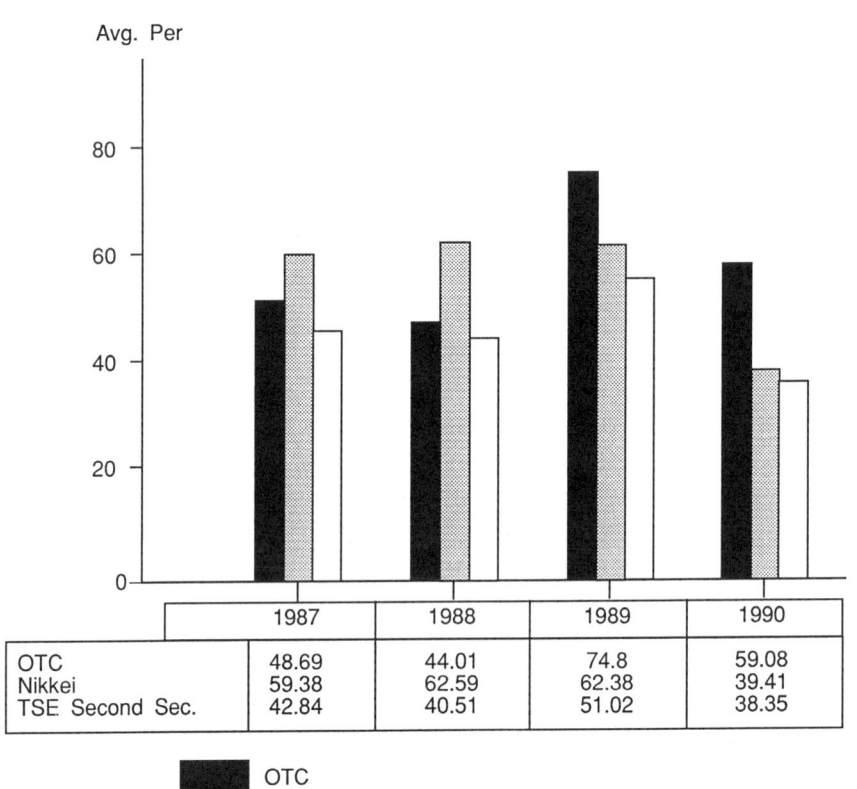

|  | 1987 | 1988 | 1989 | 1990 |
|---|---|---|---|---|
| OTC | 48.69 | 44.01 | 74.8 | 59.08 |
| Nikkei | 59.38 | 62.59 | 62.38 | 39.41 |
| TSE Second Sec. | 42.84 | 40.51 | 51.02 | 38.35 |

■ OTC
▒ Nikkei
□ TSE Second Sec.

The Nikko Research Center

## Figure 7.10

|  | NASDAQ | | Japanese OTC | |
|---|---|---|---|---|
|  | No. of Co.'s | Trading Volume (mil. shs.) | No. of Co.'s | Trading Volume (mil. shs.) |
| 1984 | 4,097 | 15,159 | 116 | 102.0 |
| 1985 | 4,136 | 20,699 | 127 | 118.2 |
| 1986 | 4,417 | 28,737 | 140 | 203.7 |
| 1987 | 4,706 | 37,890 | 151 | 181.4 |
| 1988 | 4,451 | 31,070 | 196 | 290.1 |
| 1989 | 4,293 | 33,530 | 263 | 529.2 |
| 1990 | 4,131 | 33,379 | 342 | 1,065.5 |

Nikko Securities Co. Int'l, Inc.

compared to the 342 on the Japanese OTC market. Because of this, there is a distinct difference in the trading volume between the two markets. (See Figure 7.10.)

In 1990, the trading volume on NASDAQ was over 33 billion shares (33.4 billion) compared to Japan's OTC market that was 1.1 billion shares. But one needs to keep in mind that Japan's OTC market is still relatively young compared to the one in the U.S. The substantial larger trading volume on NASDAQ is explained by the fact that the NASDAQ has a more developed quotation trading system and a greater number of traded companies.

In 1990, however, the trading volume on Japan's OTC market dramatically increased from 1989. The trading volume on the NASDAQ market, however, dramatically increased from 1989 to 1990. (See Figure 7.10.)

Historically, an explanation for this may be the number of market makers per company between the two markets.

Since 1987, NASDAQ has had an average of eight market makers per company. In the past, the Japanese OTC market had only an average of about two market makers per company. And most of the newly registered companies in 1990 had about seven market makers per company. This increase in the number of market makers has enhanced the efficiency and liquidity of this market.

## Pricing on the Japanese OTC Market

This brings us to the issue of pricing on the OTC market. Prices on the OTC market may be more volatile than those on the larger organized exchanges. Prices are subject to more volatility for three reasons: (1) the type of the issues traded on the OTC market, (2) the market making system and (3) the thin trading value. In terms of spreading value, the bid-ask spreads may be wider for the OTC market than the spreads quoted on the stock exchanges. This is largely due to the characteristic market making system of the OTC market.

Furthermore, because the roles of market makers are voluntary, liquidity in the OTC market may be lower than that of the stock exchanges. As such, it is generally accepted that stock exchanges have greater depth and liquidity than the OTC markets.

But one of the reasons why Japanese OTC stock are thinly traded is due to the type or its investors. Many investors characteristically are long-term shareholders and because of this, OTC stocks are thinly traded.

Japanese OTC stocks are less liquid than the stock exchanges because the markets lack a sophisticated computer trading system. However, as I mentioned earlier, JSDA implemented a new automatic computer system for stock trading called JASDAQ that will mirror the U.S. NASDAQ system.

This newly developed sophisticated system will improve the current QUICK computerized system that displays bid-ask prices. The QUICK system does not provide continuous market information for OTC stocks as well as information about the market makers as is the case with NASDAQ. For example, the bid-ask spreads displayed on the QUICK system do not necessarily reflect the last price at which a stock traded.

JASDAQ also displays market information for both equities and convertible bonds and will be available to both domestic and international investors.

The JASDAQ system will undoubtedly have a major impact on the OTC market by improving liquidity and pricing in the secondary market.

An important feature of Japan's securities markets are the underwriters involved in initial public offerings.

## The Role of Underwriters in the IPO Market

The OTC market has become a marketplace for growth-oriented companies to register and trade their stock. Since these companies are too small to meet the more stringent stock exchange listing criteria, the OTC market has become an important stepping stone for listing later on the stock exchanges.

As a result of this, Japan's security houses have become a major force in underwriting the domestic financing of companies seeking to raise funds in the primary capital market. The Big Four Japanese securities companies, Nikko, Nomura, Daiwa and Yamaichi, are primarily responsible for aggressively promoting initial public offerings as well as introducing other equity-related financing techniques to raise funds in Japan.

During the 1980s, the number of Japanese companies going public decreased dramatically. Along with this, the responsibilities of the securities companies promoting IPOs have also increased. These responsibilities include: (1) advising companies on forming their capital structure policies, (2) assisting in the preparation of the required documents, (3) handling negotiations with the Ministry of Finance, Japan's Securities Dealers Association and other authorities, (4) setting the issuer's initial offering price and (5) marketing and distributing new issues and public offerings to investors in Japan as well as abroad.

In view of this, guidelines regarding Underwriter Agreements between the issuing company and the underwriters have been established. The cost estimates for an IPO and OTC registration have also been standardized. For example, the underwriting fee for an OTC registration, which is the same fee for listing on the TSE, is 3.1 percent of the aggregate issue amount plus two yen per share. The underwriters also have commitments and responsibilities to the government authorities to ensure the protection of investors.

One of these responsibilities is setting the public offering price of newly issued shares or existing shares. An auction system is now used to establish the offering price for an IPO. The offering price is calculated using net earnings and net assets of comparable companies. The price calculated is the minimum price. The balance of the shares will be priced at the weighted average of the bid prices.

Although the underwriting securities companies participate in establishing the IPO price using the auction system, all other securities companies are prohibited from participating in the public auction. In addition, to protect the investors, special interest groups, the top 10 shareholders of the issuer and employees are also prohibited from participating in the public auction.

After the initial public offering, the underwriting securities are responsible for making a market in the new issues. As I mentioned earlier, the number of market makers has increased for the OTC market. This coincides with the increasing number of IPOs in Japan and its relationship to the OTC market.

## Japan Issuers of IPOs and the OTC Market

Since the mid-1980s, investment trends have been shifting from larger mature companies to growth-oriented companies. Because of this, the OTC market has grown significantly during the past few years primarily due to the increase in the number of IPOs in the Japanese capital market.

A number of factors have contributed to this increase, some of which I have mentioned earlier. These factors include: (1) improving government measures such as an easing of issuing standards and the relaxation of rules regulating the raising of capital, (2) improving market conditions such as increases in stock prices and (3) the enhanced investor confidence and image in companies when they go public.

Companies that participate in IPOs and register on the OTC market specifically consist of emerging growth companies, many of which may list on the larger stock exchanges of the future.

At a time when IPOs have become popular, companies have chosen to register on the OTC market overall because of the relaxation in registration requirements in comparison to the stock exchanges. The criteria to register with this market only requires that an IPO candidate have over 200 shareholders, a minimum net worth of $1 million, positive earnings in the latest fiscal year and two years of unqualified audited financial statements. In addition, there are no divided requirements for OTC registered companies.

Furthermore, the number of shares to be offered for public auction at the time of an OTC registration is determined by the following criteria: the number of shares outstanding times 3.75 percent plus, 175,000 shares.

In view of these increasingly relaxed requirements, more and more types of emerging companies are entering the IPO market. During the past year, the largest single group of OTC companies were commercial companies, machinery manufacturers and electronics makers.

Moreover, if we break down the OTC companies into the larger categories of manufacturing and non-manufacturing, we see that the number of non-manufacturing companies has grown in the past few years. Currently, over 50% of the OTC market consists of these types of companies compared to about 32% on the Second Section of the TSE.

## Types and Outlook of OTC Market Investors

Investors of these OTC companies consist primarily of individual investors. During the past year, individual investors accounted for 53% of the total of OTC stock investments. (See Figure 7.11.) The next largest category was corporations at 21%, followed by banks at 12%, and foreigners at 9%. Investment trusts and insurance companies combined accounted for only 5% of OTC investments. (See Figure 7.11.)

**Figure 7.11 OTC Market Investors 9/90–3/91**

Insurance Companies 1%
Banks 12%
Corporations 21%
Investment Trusts 4%
Foreign 9%
Individuals 53%

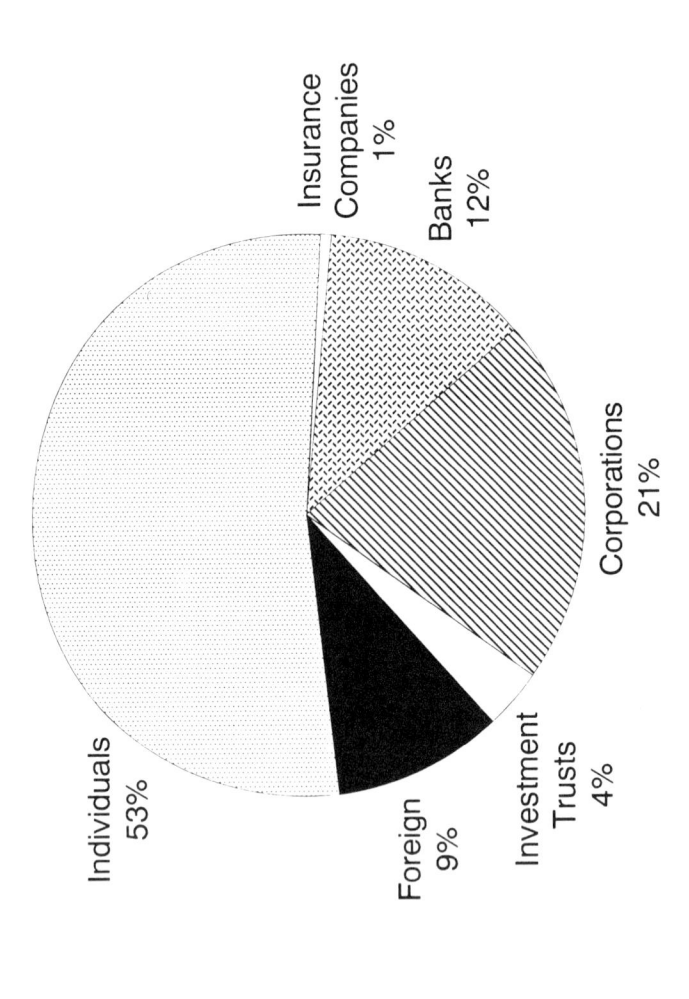

The Nikko Research Center

The outlook for demand in smaller growth-oriented stocks will continue to look good as a broader class of investors are attracted to the OTC market. Recently, increased demand for OTC stocks by institutional investors such as pension funds has encouraged more issuers to enter the IPO market.

While more investors such as institutional investors are attracted to the OTC market one can see why both the trade and turnover ratios are approaching levels seen on the Second Section of the TSE.

Influenced by the high-growth potential of OTC registered companies and preferred treatment for shareholders, investors have come to think highly of the OTC market, and because of this, it has entered a new high-growth phase. This expansion is expected to attract more foreign investment in Japan's OTC market as well.

## Why New Issuers Go Public and Outlook of the IPO and OTC Markets

Overall, the benefits of going public by a Japanese company include: (1) attractive IPO prices, (2) the realization of assets previously undervalued on the balance sheet, (3) strengthening of the corporate balance sheet by expanding access of equity financing and broadening sources of capital and (4) enhanced visibility, such as providing greater opportunity to market products. More recently, the cost of funds from equity financing in Japan has been relatively low, in comparison to other types of financing such as yen-based bank financing.

And as the Japanese OTC market experiences rapid expansion, more of Japan's own privately held companies are expected to go public, reviving what used to be a relatively stagnant initial public offering market before the 1980s

Furthermore, while Japan is experiencing more IPOs and more of these IPOs are turning to the OTC market, more foreign companies are realizing the attractiveness of raising funds in Japan and the importance of the OTC market.

# Foreign IPOs and Their Outlook on the OTC Market

According to recent data, more Japanese subsidiaries of U.S. and other foreign companies are raising capital in Japan by selling off minority stakes to the public. The growth in the number of IPOs combined with the interests in foreign equities has created an attractive opportunity to raise funds through IPOs in the Japanese market.

Between 1986 and February 1991, there were 11 initial public offerings of foreign companies' Japanese subsidiaries, eight of which are wholly owned or joint ventures of U.S. companies. (See Figure 7.12.) The total amount of funds raised by these 11 initial public offerings was $1.29 billion.

One of the reasons why foreign companies choose to sell minority interests in their Japanese subsidiaries are the relatively high-market valuations achieved in the Japanese market.

The average price earnings for these 11 foreign IPOs was 56 at the time of their public offerings. The P/E ratios ranged from as low as 25 for Shaklee Japan (IPO in July 1986) to as high as 100 for Nippon Avionics (IPO in January 1988).

As of April 4, 1991, the average price earnings ratio for the same companies similarly was 57, and ranged from 28 (Avon Products) to as high as 139 (Iwaki Glass).

Out of the eleven foreign IPOs, eight of them are registered on the OTC market as opposed to being listed on the Second Section of the TSE. Overall, the Japanese subsidiaries of foreign companies chose to register on the OTC market because it was an ideal forum for small- and medium-sized companies in Japan to raise equity. IPOs by these companies provided an opportunity to capitalize on the higher valuations in Japan.

While a Japanese subsidiary of a foreign company can derive the same benefits as a Japanese company by going public, there are additional advantages for creating a public interest in a Japanese subsidiary. These include the positive effect on the parent company's stock price through public valuation of the Japanese subsidiary and the repatriation of proceeds to the home country free of Japanese income tax.

## Figure 7.12
### Foreign Company Subsidiary/Joint Venture IPOs

| Date | Company (Parent(s)) | Funds Raised ($MM) ($ Sold To Public) | IPO P/E Multiple | Registration | Lead | Nikko Underwriting |
|---|---|---|---|---|---|---|
| Aug. 86 | Shaklee Japan (Shaklee) | $80 (19.1) | 25.4 | OTC | Nikko | 85% |
| Dec. 87 | BR31 Ice Cream (Fujiya, Baskin Robbins) | $36 (14.2) | 37.7 | OTC | Nikko | 48% |
| Dec. 87 | Avon Products (Avon) | $209 (39.8) | 67.8 | OTC | Nomura | 10% |
| Feb. 88 | Nippon Avionics (NEC, Hughes Aircraft) | $73 (25.1) | 100.0 | TSE | Daiwa | 12% |
| Oct. 88 | Iwaki Glass Company (Asahi Glass, Corning) | $187 (26.9) | 75.8 | TSE | Yamaichi | 5% |
| June 89 | Levi-Strauss Japan (Levi-Strauss) | $78 (15.7) | 38.3 | OTC | Nikko | 77% |
| Aug. 89 | Getz Brothers (Getz Brothers (Marmon Group)) | $74 (14.5) | 43.4 | OTC | Nikko | 75% |
| Sep. 89 | N. E. Chemcat (Sumitomo Metal Mining, Englehard) | $94 (14.5) | 53.9 | OTC | Daiwa | 0% |
| Jan. 90 | Memorex Telex Japan (Memorex Telex, Kanematsu) | $22 (13.9) | 52.5 | OTC | Nomura | 5% |
| Aug. 90 | Kentucky Fried Chicken Japan (Mitsubishi, Pepsico) | $339 (33.6) | 71.0 | TSE | Nikko | 50% |
| Feb. 91 | Nemic Lambda (Lambda Electronics) | $48 (14.2) | 108.4 | OTC | Yamaichi | 11% |
| Apr. 91 | Amway Japan (Amway Corporation) | $357 (7.5) | 44.5 | OTC | Yamaichi | 10% |

Foreign companies have the opportunity of going public in Japan by either selling a minority interest in a wholly owned subsidiary or in a joint venture with a Japanese company. Four of the foreign IPOs were wholly owned foreign subsidiaries. These included: Shaklee Japan, Avon Products, Levi-Strauss and Getz Brothers, all foreign subsidiaries of American companies.

The first initial public offering of a company in Japan was Shaklee Japan. In July 1986, Shaklee sold about 19% of its subsidiary and successfully raised about $82 million. Underwritten by Nikko Securities, the stock jumped from an initial offering price of 3,900 yen (about $29) to 5,700 yen (about $42) on its initial trading day.

Levi-Strauss is another example of a successful IPO by a foreign owned parent company. In June 1989, Levi-Strauss raised a total of $81 million by selling 16% of its Japanese unit. Although the parent company is a private company in the U.S., it chose to go public in Japan because of the increased prestige and public awareness public companies have in Japan.

Levi-Strauss was also the first foreign IPO in Japan to use the "auctioning price system" for IPOs. Lead managed by underwriters Nikko Securities and Goldman Sachs, the successful high IPO price was based on the price book values and price earnings ratios of comparable companies. The planned minimum price was originally Yen 2,560 but the actual auction price was Yen 2,656 (about $19).

The most recent OTC IPO by an American parent company was Amway Japan. On April 19th, 1991, the Japanese subsidiary of the Amway Corporation in the U.S., a direct marketing company, offered its shares publicly. It is estimated that Amway raised about $357 million with its IPO by offering 7.6 million shares.

Most foreign IPOs in Japan have been joint ventures with Japanese companies. Since November 1987, there have been seven such IPOs in Japan. For example, joint ventures by U.S. companies include B-R 31 Ice Cream, Nippon Avionics, Iwaki Glass and Kentucky Fried Chicken Japan. The most recent joint venture was Nemic-Lambda Co. Ltd., an electronic parts manufacturer.

Nemic-Lambda raised approximately $50 million in February 1991, by selling about 15% of its Japanese unit. The parent company, located in the U.K., decided to go public with its Japanese joint venture company for various reasons, including the opportunities for an attractive return on their Japanese investment, to raise capital locally to expand their business, and to set-up an employee stock ownership plan (ESOP) for the Japanese employees.

Finally, in deciding to go public, foreign companies need to consider how the proceeds of the IPO will be used. Foreign IPOs can offer the sale of existing shares or newly issued shares. The proceeds from the sale of existing shares may be repatriated to the subsidiary's parent company. However, the proceeds from the sale of new shares must remain in Japan. Overall, most foreign companies include the sale of primary shares to raise funds for the Japanese unit as well as offer newly issued stock to the investing public to add to the success of the IPO.

## OTC Registration Criteria and Procedures

The amount of time for an OTC candidate to register on the OTC, assuming all other criteria for registration have been completed may be as little as five months. While it is possible to do an IPO on either the TSE or the OTC market, there are two specific reasons companies choose to register on the OTC: (1) the company can retain more shares and (2) fewer procedural requirements with less stringent qualification criteria.

According to the TSE rules, as of the listing date, an applicant company should reduce the number of shares held by "special interest investors" to 80% or less of the number of shares to be listed. The OTC market only requires the company to publicly offer a minimum number of shares from the total outstanding amount. Therefore, a company registered on OTC is able to retain a larger number of shares than that of a TSE listing.

The preparation period to register on the OTC market is also shorter than listing on the TSE because it only requires two years of audited financial statements while a TSE listing calls for an applicant company to submit audited financial statements for three auditing periods.

IPO candidates for registering on the OTC market must additionally meet certain eligibility criteria including: (1) a minimum auction share price calculated by the recent auction pricing formula guidelines by the JSDA and the MOF, (2) the number of shares to be offered, (3) a guideline period for allocation of new-shares to third parties and (4) limitations on the resale of shares by third parties.

Furthermore, the documents required to register on the OTC market are classified into two categories: (1) documents required for the Registration Application to the designated regional JSDA and (2) documents required for the Security Registration with the Ministry of Finance.

Finally, a company whose shares are registered and publicly traded must comply with certain laws and regulations designed to protect the rights of the investing public. These obligations after an OTC registration include: (1) on-going document requirements concerning the financial status of the company and number of shareholders, (2) disclosure obligations as prescribed by the Securities and Exchange Law of Japan, (3) audited financial statements to be approved by Japan's accounting standards as prescribed by both the Commercial Code of Japan and Securities Exchange Law of Japan and (4) OTC requirements as regulated by JSDA concerning the availability of business results each business period to the investing public.

## Conclusions and Outlook of the OTC Market in Japan

In conclusion, the OTC market historically is small compared to Japan's stock exchanges, or even to other OTC markets like the one in the United States. But lately, due to favoring market conditions, Japan is now experiencing a boom in IPOs. This currently is emphasizing a greater importance for the OTC market.

Growth in the IPO market will also continue to be influenced by increasingly relaxed registration requirements and developments to improve liquidity and pricing on the OTC market. Plans to install the on-line computerized JASDAQ system will not only greatly improve trading in the secondary market, but it will also greatly facilitate the growing boom in IPOs by accommodating more volume and issues.

Because of the recent surge in the number of IPOs, the OTC market is approaching the size and integrity of Tokyo's Second Section stock exchange. And with its developments to improve liquidity and pricing, in just a few years the OTC market is expected to exceed the TSE Second Section in terms of the number of companies registered and total market value.

Finally, with the growing importance of the OTC market, more and more foreign companies are being attracted to Japan's IPO market. Already, at least 10 American companies are planning to take advantage of public offerings of their Japanese subsidiaries within the year.

*Mr. Hidemi Suzuki is Executive Vice-President and General Manager of the Corporate Finance Office of the Nikko Securities Co. International, Inc.*

# Chapter Eight

# Benefitting from Japanese Corporate Venture Capital: Opportunities and Challenges

### Neeraj Bhargava

The American high-technology entrepreneur has rarely been alone. In fact, in the early days of creating and building a new business, getting by with a little help from friends has been more the rule and not the exception. In recent times, the friends have often been Japanese.

MIPS Computers, Vitelic Corp., Mycogen, Liposome Company: all these are examples of companies that have had success in tapping funds from Japanese corporate investors during their start-up and growth phases. According to a Venture Economics, Inc. estimate, the

Japanese equity investment in U.S. companies was $320 million in 1989, up from $42 million in 1985.

Capital infusion is not the only form of assistance that U.S.-based start-ups have got in such link-ups with Japanese corporations. Distribution contracts, manufacturing arrangements and joint R&D ventures have often accompanied the Japanese investments, quite often providing the start-ups with the opportunities of growing rapidly and building global businesses.

The Japanese corporations are not alone causing the recent upsurge in corporate venture capital (CVC) investments in U.S.-based start-ups. European, U.S. and more recently, Korean and Taiwanese corporations have also accelerated their investments in start-ups in the last few years. Particularly striking is the case of IBM that almost every week is reporting a new minority equity investment in small, growing companies. In a recessionary period, when private venture funds and financial investors are strapped for cash, investments from corporations provide some respite to the start-up community that is desperate for funds. More than that, CVC is changing the way small companies and venture funds view their options for adding value.

## The Logic Behind Corporate Venture Capital

From a strategic perspective, CVC provides many potential benefits for both the investor and the start-up. For the corporate investor, there are four significant benefits:

- *Building new business:* Japanese corporation, Kubota is a leading example of a company specializing in machinery and industrial products using CVC to diversify into the computer industry. Similarly, Japanese chemical companies are investing heavily into the biotechnology area to pursue new markets.

- *Strengthening existing business:* An example is IBMs efforts to establish a network of companies, developing software for its products, by making minority investments.

- *Efficient utilization of existing resources:* Japanese conglomerates such as Mitsubishi and Mitsui are constantly on the look-out for new products that they can distribute through their global distribution network. In other cases, Japanese semiconductor companies are utilizing their advanced manufacturing capabilities to produce chips designed by U.S.-based semiconductor start-ups.

- *Finding more effective ways of competing:* Small companies are often more effective than large corporations in designing new products, cutting lead time to manufacturing and responding to environmental changes. Large corporations are finding tremendous benefits in outsourcing various activities in their value-chain from small start-ups and concentrating on their core businesses.

For small companies, too, the strategic benefits are no less impressive. CVC investments provide them with a considerable degree of credibility among customers, suppliers and investors. If the investments are a part of a broader alliance providing them access to their investor's marketing, R&D or manufacturing capabilities, the small companies stand to benefit even more. They have the opportunity to address larger markets and the chance to combat global competition more effectively.

CVC investments also offer opportunities to both the parties involved to create value, that might not have been possible otherwise. My own research on start-ups, in the information technology industries, launched in the 1980–85 period has shown that the stocks of the ones with significant corporate investors significantly outperform others over a five-year period. For the corporate investor also, the returns from equity investments in the stocks of start-ups significantly exceeded the returns from a comparative investment in its own stock. Divestment by the large investor has often yielded profits several times the original investment, as was observed in the cases of Apple Computers' investment in Adobe Systems, and Olivetti's in Stratus and Abbott Laboratories' in Amgen.

# Japanese Corporate Venture Capital

While value creation might not be their primary goal, Japanese corporations are very much riding the CVC wave and pursuing new opportunities in the U.S. Not only are they offering funding to the start-ups, but also an approach very different from U.S.-based investors. "The Japanese industrial corporations are willing to wait much longer for their returns. They view the American approach to corporate strategy with amazement and feel that we exert too much pressure on the start-ups to generate early returns and go public. They are more concerned with product and market supremacy in the belief that success in these areas will eventually lead to appropriate profitability," says Arthur C. Spinner, general partner of Hambro International Equity Partners, New York City.

The Japanese corporations also bring to the table a global network and quite often a brand-name with a global appeal. In the last two decades, corporations such as Mitsubishi, Matsushita, Sony, Toyota, Toshiba among others have painstakingly built a global image of quality, innovativeness and reliability. Their strengths lie not only in their technological prowess but also in their ability to distribute, quite often, a wide variety of products. They are also quite capable of leveraging these resources and creating global businesses out of the technologies offered by U.S. start-ups.

For a high-technology start-up, success in today's global economy hinges on its abilities to address as wide a market as possible and to do it fast. This is extremely vital in today's highly competitive environment when product life cycle's are getting shorter and the time available for benefitting from innovation is shrinking.

An Arthur D. Little study in 1990, of small high-technology companies, showed that the most profitable companies grow rapidly and make the most efficient use of their assets. For a small company in the U.S., often facing impatient stockholders, an alliance with a Japanese company, which is also an equity investor, is one of the few avenues for rapid growth and effective utilization of its core strengths.

## Venture Funds as Co-Investors

The challenges for the Japanese companies is identifying the most appropriate opportunities among a myriad of proposals and business plans. Investing with U.S. venture funds is an approach by setting up corporate development funds is increasingly being chosen, a leading example being Mitsui with Orien Ventures.

Another example of a fund backed by the Japanese is Hambro International's global fund, in which Mitsubishi is a key investor. In this case, however, the emphasis is not on corporate development but on building value of the portfolio companies by utilizing the global network and resources of investors such as Mitsubishi.

Such funds provide the Japanese corporation with a partner that has a better access to private equity markets, experience to recognize situations to stay away from and the opportunity to save valuable time and effort. For the venture funds, it provides a new method of adding value for their companies.

"We always look at the companies we invest in from the perspective of how we can add value to what they are already doing. Our link with Mitsubishi provides us with a new network with tremendous scope to tap global opportunities for our portfolio companies," says partner, Arthur Spinner. Co-investment also lowers the risk of the investment as the presence of large corporate backer increases the chances of the start-up's survival and profitability.

Another approach is creating a venture firm that specializes in developing international corporate partnering agreements, joint ventures, acquisitions, technology transfers and other ways to effectively invest their CVC. A leading example of such a venture firm is Boston-based Techno-Venture, that is backed by corporate investors such as Nippon Mining Co. Ltd, Hitachi Ltd., Nissan Motor Co. and Mitsui Bank.

In these approaches, the venture firms have a tough act to perform. Involving corporate funds in their investments means balancing both the strategic interests of the corporate investor and the value creation process in the small company. Not all outcomes of such investments will end up satisfying all the parties involved.

## Figure 8.1
## Corporate Venture Capital Investment Outcome Matrix

| Strategic Benefits for Corporate Investors | | Low (Value creation) | High (Value creation) |
|---|---|---|---|
| | High | Does not suit the venture fund. | Ideal conditions for both the venture fund and the corporate investor. |
| | Low | Does not suit both. | Does not suit the corporate investor. |

Value creation in the small company in which the investment is made.

All four outcomes mentioned in Figure 8.1 are possible because all aspects critical to a venture's success are not always controllable or predictable. For example, in science-related areas such as biotechnology, R&D results cannot be easily predicted. Quite often, the end product is quite different from what was planned. For the corporate investor, strategic benefits in such cases might not turn out to be what were originally planned or perceived.

Strategic benefits also depend on how well the synergies between the corporate investor and the start-up are managed over a period of time. As a start-up grows and meets some success in areas originally planned, it also develops its own strategies for growth and diversification, which might not suit the interests of the corporate investor. Managing strategic convergence and commitment toward each other over a period of time is by no means an easy job. The Apple Computers and Adobe Systems case is one example where the commitment toward each other's interests was often tough to manage over a period of time.

Another relevant issue is the degree of importance that the corporate investors attach to value creation in the small company and the return on investment. "Strategic issues and not ROI drive our corporate investments. We focus on evaluating the product being developed by the small company and its potential market. Unlike venture capital companies, our people do not attribute as much importance to evaluating the management of a start-up, although I think they should be doing more of that" says Philip N. Sussman, director of strategy and business development at Ciba Geigy's pharmaceuticals division. Also, the size of investment is often too small for corporate investors to be overly concerned about adequate financial returns.

Corporate investors, however, realize that strategic benefits are often attained over a long period of time. In this way. they are more likely to be a restraining force on the venture firm that is co-investing with them. This is often the case with the Japanese corporate investors. The venture firm could, however, have its own axe to grind. Its institutional investors might not be equally patient about returns and the fund managers, therefore, have quite a balancing act to perform.

The start-up has its own share of problems. Different stakeholders set different yardsticks for measuring its performance. Managing a variety of complex relationships becomes a significant challenge for its managers. However, they also benefit from the longer-term perspective taken by the corporate investors and it takes some pressure off them to generate quick returns.

Despite these potential problems, the rapid influx of Japanese CVC is providing both venture funds and start-ups in the U.S. tremendous opportunities to build new businesses and create value. Considering the massive financial reserves that Japanese corporations have compiled over the last decade, the capital in-flow is likely to continue in the 1990s. For the high-technology American entrepreneur in the 1990s, not only are the markets and competition global, but also the sources of funding and the availability of critical resources to build businesses.

*Neeraj Bhargava is a New York-based consultant working with high-technology industry companies on business planning and new market development. He is also the author of several cases on corporate venture capital investments.*

# Chapter Nine

# U.S.-Japan Strategic Partnership: The Use of Technology Transfer and International Network

## Dr. Yaichi Ayukawa

This article will elucidate similarities among the respective economies, governmental policies and cultural world views of the United States and Japan in addition to exploring the role of venture capital in the two countries.

Although Japan is concerned with economic problems faced by the United States today including the trade imbalance, capital deficits, unemployment and weakened industries, it is also impressed by progress American industry has made in the areas of biotechnology, microelectronics superconductivity and technical innovation.

Unlike Japan, however, the United States can be accurately described as a nation of "I's" rather than "we's." For example, the American entrepreneur with a science or engineering background is the hero on college campuses and in business circles. According to a recent survey conducted at M.I.T., 54% of the undergraduates and 49% of the graduates said it was very important for them either to be top managers in a firm or to run their own business. This mindset may be summarized by the statement: "The heroes of these students are the engineers—entrepreneurs of Route 128 and the Silicon Valley."

Japan, needless to say, is different. In the educational system, individualism, originality and entrepreneurship continue to be frowned upon. Young graduates entering the job market are as eager as their predecessors to join a big, "safe" company, and investors share a similar attitude towards their money. So, the American ace-in-the-hole in its competition with Japan may well constitute Japan's system of education. The Japanese educational system has created a highly literate, disciplined yet collective work force. And while the risk-taking entrepreneur does exist in Japan, he or she is most certainly in a minority.

Secondly, in Japan, it is very difficult to say "I." Instead, people feel compelled to refer to one another as a group, discarding first person references and expressing themselves as "we." Governmental issues are also handled in this collective way. The Japanese government declares that because the United States is putting pressure on individual automobile imports from Japan, the country, as a whole, must implement a quota system. Using the same technique, the Japanese government exhorts local farmers to band together in the face of their American agricultural counterparts' demand for the liberalization of rice imports. The indirect approach of the Japanese is the main source of U.S.-Japan miscommunication and friction. This ambiguous strategy also discourages entrepreneurs in Japan from taking the initiative, a kind of daring that is essential to real progress.

In the free U.S. society, where "individualism" is respected, a businessman who tries to realize a personal vision and make unprecedented decisions may be more likely to win the race. In Japan, on the other hand, such evidence of entrepreneurship is often held in contempt, thereby curbing potential innovation.

Certainly, this does not represent a mindless endorsement of individualism, which also has its negative aspects. For instance, the overpowered ego-hungry individualism may be the cause of many unfair practices observed in the United States such as unscrupulous takeover bids, antagonistic management buyouts and predatory head-hunting.

## Venture Capital in the United States and Japan

The role of venture capital in the United States and Japan is, clearly, affected by the respective world views of these two nations. Venture capital has played a key role in the rejuvenation of entrepreneurship in the States. There are four key elements that have inspired the emergence of this individually capitalistic economic environment:

First, America possesses a positive climate for free and private capital enterprise.

Second, after the Vietnam War, excess funds were generated and held by institutions and industries, especially bank, insurance companies, foundations, pension funds, and educational organizations, in spite of inflation and the rise of interest rates.

Third, the reduction of the capital gains tax encouraged an increase in investment.

And finally, venture capital management styles recently shifted from mere interest in the financial investment to actual participation in management. Such a change was possible because of the availability of experienced personnel endowed with both technical and business backgrounds. These "new" venture capitalists are able to act as general partners not only in screening and nurturing venture business but also in assuming responsibility for generating profits.

In Japan, however, the venture capital concept is still in a theoretical phase although there have been some success stories of entrepreneurial "venture" businesses in Japan that began several decades ago including those of Honda, Matsushita, YKK, Kyocera and Sony.

But, generally, the role of venture capital as it exists in the United States has played a very minor role in enhancing Japanese industry.

What then has spurred the impressive Japanese economic growth? Part of it is due to the fact that Japan has been acquiring advanced technologies from the United States and Europe since the end of World War II, and has improved upon them through the application of intelligence, talent, diligence and a hard work ethic. One could even say that Japan drew its inspiration from the West by efficiently making variations on already introduced techniques at a relatively low cost. With few exceptions, it can be accurately asserted that, so far, Japan's creative innovations have been based on those originated in other countries.

For example, Japan's purchase of technological information from the U.S. totalled more than $1 billion in 1988 while the United States bought less than half that amount from Japan. However, quality in the future cannot be maintained exclusively by adaptation of borrowed technologies. The United States and Europe are likely to impose more protective mechanisms to prevent other nations, like Japan, from borrowing their advanced technologies.

Given these prospects, Japan's most reasonable alternative would be to develop its own technologies from the inception stage. And an economic environment conducive to private venture capital would be essential in achieving that goal.

Why then has such an entrepreneurial climate not emerged until this point and what can be done to ensure that it does materialize in the future?

The main reason that Japan lags behind the United States in venture capital is that the total accumulation in absolute terms is still less than in the United States, even with Japan's current economic strength. Following the enforcement of the Anti-trust and Economic Decentralization Laws and the abolition of holding companies after World War II, Japanese industrial development depended on financial rather than industrial capital. Furthermore, Japan still prohibits the holding company system even in this period of internationalization while the holding of a company's own shares is also limited by law. Therefore, in Japan, it is rather difficult to carry out mergers and acquisitions through share manipulation. Banks are more reluctant to invest in still unproven ideas and new businesses, while most entrepreneurs try to succeed within the corporate structure. These conditions make it very difficult for an individual or private group with a venture business idea to accumulate the necessary capital. Hence, new business and technical developments have largely come about through internal ventures within large, existing corporations.

A second reason is that the over-the-counter, (OTC) stock market for venture-backed firms is highly developed in the United States. But, in Japan, the OTC market is still highly regulated. At year end 1989, only some 260 companies were listed on the OTC as compared to 4,300 on NASDAQ. The Japanese management and the security firms must maintain a conservative attitude with a strong sense of responsibility toward public investors who might risk their funds in a new business upon public offering of stocks. However, there is a trend to encourage some relaxation of the restrictions. A special countermeasure to reduce the capital gains tax, as was achieved in the United States, is being requested of the government, since this tax currently amounts to between 50% and 70% in Japan.

The number of newly established ventures in the United States in fiscal year 1975 was 326 firms, compared to only 95 in Japan, while in fiscal year 1986, there were 702 new start-ups in the United States and only 104 in Japan, nearly 1/7, indicating a much faster pace in the growth of the U.S. ventures.

On NASDAQ, 40% of the companies took only up to 10 years to reach the IPO stage, whereas on the Japanese OTC 80% required as long as 15–50 years.

The total venture capital committed in the United States in 1988 was $30 billion as compared to only $2.6 billion (1/11th) in Japan.

Investments in start-ups are not the sole goal of a venture capital company. Since most venture capital firms in Japan are linked to a bank or a securities company, they often serve more as a means of developing new clients. Loans, factoring and leasing carry more weight than straight investment. Japanese venture capital firms were seen to prefer lower risk deals compared to their American counterparts. Japanese investors' lack of active involvement in the growth and management of start-up businesses contrasts sharply with the American situation. An American venture capitalist most often actively provides whatever expertise has to offer, which is gladly accepted by the entrepreneur, who would rather focus his or talents on creation and innovation. In Japan, however, not only must FTC restrictions be surmounted but also a dire situation is required before the investor steps in.

The Japanese venture capital industry may be described as being in its incipient stages. While the younger Japanese may be personally suited to venture business, the Japanese economy, in its drive to become the world's most efficient mass producer of consumer and industrial goods, has left little room for small, independent companies to make a significant contribution.

Scott McNealey, C.E.O. and founder of Sun Microsystems, recently gave a barn-burning review of Sun's successful venture, and captivated the audience with the exciting story and his sharp style. At the end of his speech, there were some typically timid questions from the Japanese audience, such as "What advice would you give to Japanese companies?" "What does it take to be a successful entrepreneur?" and so on. In answering, he listed the following points without hesitation:

1. Make fast decisions

2. Defy conventional wisdom

3. Make major conceptional leaps but not incremental changes

4. Embrace unproven ideas and technology

5. Take big risks

In listening to Scott's reply there were a few gasps, and then deafening silence. All the top managers present acknowledged that this recipe has worked well in the American Society, but they could not see it work in Japan. Scott's comments and the Japanese reactions are a typical example of the basic differences in the respective cultural backgrounds and the structure of the two societies.

Generally speaking, within the Japanese business society,

1.  Consensus stifles quick decision making

2.  Risk-takers are treated with caution

3.  Entrepreneurs are often regarded as outcasts

4.  Top managers are risk-averse, rigid and slow decision makers

In Japan, multidisciplinary experts are rare. Moreover, the idea of "venture" is scary—the popular image of this concept emphasizes the risk element, rather than the idea of a researched investment. Japanese society and businesses are vertically structured. Consensus and team efforts are prized. A genius' brilliance can be quickly smothered in a climate controlled be a group of average men striving for harmony. Japan today may be a land of many prosperous and good things, but it is not a fertile ground for entrepreneurial activities.

However, all is not so grim. It is possible for one to say that Japan may be heralding its own "Perestroika" vis-à-vis the idea of venture and enterprise. Some encouraging signs are apparent.

First, restructuring is forcing many firms in mature or saturated markets to diversify into new areas, including service industries and high technology. Giant Nippon Steel, for example, the world's largest steel anchor, is pursuing an aggressive diversification program to reduce dependence on its traditional business. So far it has established a few joint ventures, one of which is a computer software company in which Techno-Venture has invested.

The impact of the global stock market crash—Black Monday—is still being sorted out. Scandals involving possible price fixing, insider trading—as demonstrated in the case of "Recruit Cosmo"—and others shook the government and the OTC market and made it an uncertain bet at best for IPOs in the near future. The OTC market, however, may become increasingly attractive to small- and medium-size investors, if the Tokyo Stock Exchange climbs to unaffordable heights. The government's continued program of financial liberalization and deregulation represents venture capital's greatest hope. If taxes and financial regulations are significantly liberalized, a U.S.-style venture capital industry might have a chance to evolve.

Fortunately, there appear to be some signs that further liberalization will occur in due course. There is also a large amount of excess capital beginning to accumulate in Japan. Another factor that will contribute to additional venture capital activity is the large number of independently-wealthy, private individuals who are starting to be involved in the venture business. Thus, they offer better risk profiles than well-structured institutions. This trend may also see a decline in the importance of the life-long employment system.

Another factor that will have an impact on the development of the venture capital community in Japan is the process of globalization. A decade ago, Japan was not aware of what was going on in the U.S. venture capital market. Today, the Japanese as well as others around the world are instantly aware of developments in the international venture capital markets. While Japan's venture capital industry has undergone many ups and downs, I believe its future role in developing new business will continue to grow in importance.

It has been recently pointed out that Japan is scouting U.S. labs and small firms for seeds of new products; Japan is harvesting the creativity of the American mind. But the same can be said of the United States. Japan should be proud that MIT pioneered in setting up its Industrial Liaison Program office in Japan some 12 years ago and currently has a very successful program involving some 50 top corporate members. MIT has raised a substantial amount of money in endowments from Japan so far and is also enjoying a bilateral flow of knowledge between MIT scholars and Japanese industrial society.

MIT is also making a strong impact in Japanese industrial circles by offering the well-managed Sloan School program, that teaches international and multidiciplined management skills to a very select Japanese business elite. Japanese venture capitalists hope for the gradual emergence of more management people, who not only have extensive knowledge about the finance and security businesses but also are capable of understanding interdisciplinary technologies.

Even if to a lesser extent than in the United States, Japan obviously has capital and people who are willing to invest in it, and it has an educated and talented population. A catalyst or a stage-setter is needed now.

For example, in the Kabuki Theatre, one of the most important characters is the Kuroko, the person dressed in black who appears before every scene to introduce the plot. His or her role is crucial because he or she has to set the stage—where all of the elements are brought together to bring about a favorable reaction, a desired result. The venture concept in Japan needs such a Kuroko, or stage-setter.

## Techno-Venture and Its Role in Japan and the United States

As one of the pioneering and independent venture capitalists in Japan, originating from an all-around management service company entering into international venture investment activities, Techno-Venture (T-V) has been exerting its best efforts in creating a value-added bridge for U.S.-Japan venture business partnerships with technology, capital and management.

Techno-Venture is fortunate to have special venture capital ties with two leading firms; Kleiner, Perkins, Caufild & Byers in San Francisco and Advent International/TA Associates in Boston.

Techno-Venture has thus far created five partnership funds including both major Japanese financial institutions (Nippon Life Insurance, The Bank of Tokyo, Mitsui Bank, Long-Term Credit Banks) and corporate partners (Nippon Mining, Nissan Motors, Mitsubishi Corp., Ajinomoto) and foreign investors (Harvard Management, Credit Lyonnais, Wheat First Securities, etc.) totalling $200MM.

A large portion of Techno-Venture's funds is invested in U.S. ventures with the remainder invested in Japan. Once Techno-Venture has invested in and assisted those U.S. ventures until they have reached a certain stage of maturity, then it helps them come into Japan, as licensing joint ventures or contractual R&Ds,—depending on the specific orientation of their business. In most cases, Techno-Venture also provides comprehensive business support in the form of management, top executive recruiting, office provision, contractual relations with governmental agencies and business partners, as well as local financing.

As an example, Techno-Venture has a management service contract with such high-tech outfits as Genentech. In the early phase, it arranged for Japanese investors to buy Genentech stock in the United States through private placement. Later, it assisted in the formation of strategies for R&D and commercial partnering transactions with leading Japanese pharmaceutical companies. T-V also assisted in establishing Genentech's own subsidiary—Genentech K.K., thereby creating a true expansion program.

In another case, T-V invested through a fund, managed by the highest return earner VC KPCB, in the Liposome Co. (TLC), a New Jersey-based, very sophisticated phoso-lipid drug delivery system company, and later represented a $5MM syndicated direct investment. Subsequently, the Liposome Company IPOed in the United States. In step wise approaches, in the first phase, Techno-Venture conducted market and technical surveys for TLCs entry into Japan. After successful conclusion of R&D contractual arrangements with major pharmaceutical companies in Japan, it also was instrumental in executing a technology transfer of TLCs basic lipid technology to Nippon Oil & Fats (NOF), a leading Japanese chemical company, as well as a stockholder of T-Vs.

In Phase III, Techno-Venture assisted in the establishment of a joint venture among The Liposome Co., Nippon Oil and Fats for the exclusive supply and marketing of these companies lipo products to their R&D partner drug companies.

Now in Phase IV, Nippon Oil and Fats further invested in the Liposome Co. so that they could jointly expand sales/marketing and improve engineering/production capabilities world-wide. This is a typical case of how it acts as not only an ordinary VC but also provides various value-added management services. As a result of its services, it had a premium buy-back arrangement with TLC/NOF for invested equity to obtain a capital gain instead of going through a long and difficult IPO process.

Techno-Venture also made similar deals in the computer business area by investing in promising U.S. computer-related ventures, screened and matured by U.S. VC partners in the early stages. It

## U.S.-Japan Strategic Partnership 131

assisted in establishing their subsidiaries or joint ventures in Japan by providing support services. These include recruiting Dr. Amo, of Toshiba, to be President of Sun Micro Japan. We, also, have made sequential investments in the Japanese subsidiaries, and helped open doors through contacts with the top management of various clients in Japan, as with the U.S.-based computer companies, Nihon Alliant and Nihon Sun Micro Systems.

Another noteworthy case is Aspen Technology, Inc, which provides unique process simulation software to both the chemical and bio industries. There was a feature story about MIT recently televised by NHK and Aspen Technology headed by Professor Larry Evans, who was mentioned as a success in the spin-off of technology form MIT. Jointly with our U.S. partner, Advent International, Techno-Venture assisted Aspen Technology in establishing its overseas leadership by investing in the U.S parent and setting-up a Japanese branch. With help and through a well-balanced U.S.-Japan working team, Aspen Technology successfully captured major Japanese clients in a short time period. Given the difficulties and competitiveness of the market, the achievements of Aspen Technology is indeed, significant. There are now efforts to establish this company's subsidiary to make a long lasting and strong bridgehead in Japan.

It is commonly believed that a corporate takeover within Japan is regarded as taboo or even considered as high-jacking. But there are signs that attitudes are changing as managers develop a taste for buying foreign companies.

According to a recent Japan Economic Industry Journal opinion poll of some 600 Japanese executives, two-thirds were "interested" in M&A (mergers and acquisitions). Even today when Japanese firms talk of M&A, they mean overseas purchases. In a country where top companies offer their employees lifetime employment, buying another firm is seen as the exception. Corporate culture is so strong and varies so significantly in Japan that the management of two different firms is usually too painful to contemplate.

However, acquisition activity in Japan has taken place in distinct waves since 1945. In the first decade or so after the war, it hinged around the banks, retailing, and pharmaceutical, many of which reasserted control over groups that were broken up by the U.S.

occupation government. A more recent wave has coincided with the rise of the yen, coupled with the strategic move of diversification and improvement of companies suffering from a structured recession that forced many large Japanese companies to make a move towards M&A activities.

M&A is going to be a rather important means for providing liquidity of VC investments. Techno-Venture has acted as intermediary in bringing about a unique M&A project by which Misawa Homes acquired a TSE Section I listed company (Nippon Eternity Pipe Co., Ltd., formerly owned by Nippon Cement) and have since converted it into a value-added housing and resort business.

Further, under a newly initiated approach, Techno-V USA Limited Partnership, called the Primer Fund has been established to prime international "Corporate Partnering" transactions. This Fund establishes bridges between Japan and the United States by investment participation, in the form of a minority position, in a particular company, which can be either Japanese or American, on behalf of a potential buyer or seller.

The Primer Fund permits a "courtship period" during which the two companies can get acquainted. It also offers the possibility of priming and inducing a larger flow of capital while maintaining the continuity of management.

The Fund makes a prior arrangement to transfer its minority interest to either or both partners at a pre-set time and price. This avoids the obvious pitfalls of the present type of instantaneous and possible hostile M&A.

One of the other approaches is to identify high-potential ventures in Japan. The investment and development is carried out jointly by the U.S.-Japan corporate partners, and then the technologies or products are brought to the United States by establishing a U.S. subsidiary of the Japanese joint venture. Techno-Venture's objective is to harness foreign ideas and inventions with Japan's talent for adaptation and modification to produce improved secondary and tertiary products.

The developed technology can be implemented not only in Japan but also back in the foreign countries—a two-way flow. This will achieve even greater financial returns through not only capital flow, but also technology and product flow. Incidentally, while working

## U.S.-Japan Strategic Partnership 133

for MITI as a member of the executive advisory board for the planning and setting up of the infrastructure of new R&D policies, formulated certain concepts to improve U.S.-Japan-ASEAN relationships by geographically circumscribing the major elements of business development, namely research, development, production and marketing. The concept works as follows.

Generally speaking, the United States is great in the creation of new ideas and technology and perhaps research. However, when it comes to production and marketing, Japan is surpassing the United States, and the ASEAN countries are growing stronger in effective production. The venture capitalist is best situated to harness these "talents" by "creating" a system for various nations and entities within nations, to take advantage of their respective expertise.

With this idea, together, with Advent International, Techno-Venture established a system called AMMI (Advent Manufacturing/Marketing International,) whereby the United States offers the results of research, Japan carries out development, Singapore or Hong Kong handle production, and finally the United States and Japan provide the marketing opportunity.

For example, through Techno-Venture's introduction and investment, a Singapore manufacturer has recently initiated a manufacturing project for high quality video tape through licensing arrangements with a leading Japanese video tape system maker. The finished product is to be marketed in the United States and Japan via a newly formulated marketing organization through T-V's investment. This system has wide application because it is based on the truism that while there are differences between two nations, the business goals and desires are fundamentally the same.

If we look at the long range co-existence of the United States and Japan, we should be working to have the United States regain its past superiority in mass-production/QC skills while Japan should work harder to develop its creativity and basic R&D capabilities. This approach will ensure independence, while allowing them both to offer and enjoy very lucrative markets together.

It is fortuitous that MIT took the initiative in setting-up a U.S. Productivity Study Commission to review and suggest ways to regain the recognition the United States had once maintained for its productivity skills. The representatives of the Commission have recently visited Japan. In response to such U.S. endeavors, Japanese leaders at both the private and governmental levels are also very eager to organize a counterpart study in Japan to work out the two nation's problems hand-in-hand. The U.S. study team should further extend its research to an analysis of U.S. companies in Japan, and a Japanese study team should conduct similar inquiries in the United States.

Incidentally, both nations are placing too much of an emphasis on the issues of national boundaries; that is, trade and capital imbalances and exchange rates. They are also overscrutinizing such flows over international borders.

If we closely check the hidden figures such as sales and profit and investments made by U.S. companies in Japan (IBM, Intel, Motorola, Xerox, etc.) and likewise, the counterparts provided by Japanese firms present in the United States (Toyota, Nissan, Honda, YKK, Sony, etc., etc.), it appears that not much of an imbalance exists.

According to Kenichi Ohmae, of McKinsey, Japan imports products worth $25.6 billion and buys made-in-Japan American goods valuing $43.9 billion a year. In total, American companies have $69.5 billion worth of products in Japan. On the other hand, the United States imports $56.8 billion worth of goods from Japan and buys $12.8 billion goods manufactured and sold in the United States by Japanese companies, totalling $69.6 billion—which is almost equal to that of Japan.

In other words, the market penetration of both countries into each other's turf is not even nearly identical. Lack of awareness of these statistics coupled with the worst public relations on the part of the Japanese has uncertainty among the American public.

We all agree that on the surface Japan is doing very well at present, with high productivity, low inflation, high savings rates, and a positive balance of payments. However, it is difficult to be equally optimistic about the future state of the Japanese economy.

The recent volatility in Japan's financial markets has been anticipated. In the past several weeks the TSE, the bond market, and the yen have all tumbled. Recently the Nikkei index dropped as low as 33,321, or 14% from its record high of 38,712 in early January 1990. And the yen's value continued to decrease, closing at nearly 150 yen to the U.S. dollar in the last few days. Although the double threat of rising interest rates and inflation are probably key reasons behind these recent declines, the market disturbances also reflect other weaknesses in the Japanese economy.

First, Japan relies heavily on foreign suppliers for natural resources. In particular, satisfaction of Japan's vast energy needs is dependent on imports of oil, the price of which has continued to creep upward in the past two years. Second, Japan's real estate values, especially in metropolitan districts such as Tokyo, are highly overinflated. This overvaluation has led to inflated asset appraisals of companies and encouraged the flow of capital into "zaiteku" (or capital management) services and leisure consumption businesses instead of the manufacturing sector.

One consequence of this redirection of capital flow is that consumer debt financing has become common practice in Japan, similar to the situation 30 years ago in the United States. Given the country's lack of natural resources, it will be quite damaging if Japan follows the path of U.S. consumerism and does pay sufficient attention to manufacturing productivity. If this trend were to continue, it would indeed be a fearful future.

Japanese leaders certainly ought to carefully weigh these points and respond accordingly before it is too late. At the same time, Japan should take the initiative in offering to its advantages to other nations, namely, skills in labor management, product innovation, etc.

In conclusion, differences in management styles between the United States and Japan present both opportunities and difficulties. However, by realizing these differences, each respective management team can make progress.

Mutual cooperation is a positive way of alleviating the economic friction that has been allowed to develop and a means of creating an interdependent alliance that will prove to be prosperous for both nations.

*Dr. Yaichi Ayukawa is a founder of Techno-Venture Co., Ltd., and serves as its president and CEO.*

# Chapter Ten

# Entrepreneurs in Japan and Silicon Valley: A Study of Perceived Differences

Takeru Ohe, Shuji Honjo, Mark Oliva and Ian C. MacMillan

This paper presents the preliminary and presently speculative conclusions of a psychological study of entrepreneurial phenomena in Japan and Silicon Valley. A questionnaire was developed to identify

Reprinted by permission of the publisher from "Entrepreneurs in Japan and Silicon Valley: A Study of Perceived Differences," by Takeru Ohe, Shuji Honjo, Mark Oliva and Ian C. MacMillan, *Journal of Business Venturing* 6, No. 2, pp. 135–144. Copyright © 1991 by Elsevier Science Publishing Co., Inc.

two major ways in which entrepreneurs were different from average managers of large corporations. First, the entrepreneurs' perceived difference between themselves and managers was measured to create a Personal Difference Index (PDI). Second, the entrepreneurs' perceived difference between their firms and a typical large firm was measured to develop a Corporate Difference Index (CDI).

Our primary finding is that U.S. high-tech and Japanese entrepreneurs have the same minimum hurdle degree of entrepreneurial spirit. Both U.S. and Japanese entrepreneurs require a certain minimum personal and corporate difference to overcome the obstacles to becoming an entrepreneur. However, the types of entrepreneurs in Japan and Silicon Valley are different. We also found that entrepreneurs with higher growth firms fell within a certain range of PDIs and CDIs. Entrepreneurs or entrepreneurial firms that were too similar or too different from corporate counterparts tended to fail or to remain small.

This report presents the preliminary and presently speculative conclusions from a study of the characteristics of entrepreneurs in Japan and Silicon Valley. As a result of this study, we hope to understand better the following:

- What are the universal characteristics of entrepreneurs?

- Is it possible to identify successful entrepreneurs and ventures?

- What is the effect of environment on entrepreneurs?

Our previous study[1] showed that Japanese entrepreneurs have strongly perceived differences and appropriate balance between personal and corporate identities that is reflected in similar levels of PDI and CDI. Entrepreneurship is a rare career choice in Japan, where the general pattern is to aspire to an executive position in a large company. The successful Japanese executive seeks to suppress his individual aspirations and to conform to the norms of the homogeneous groups with which he works. In the United States, however, individual identities are strongly encouraged, creating a strong entrepreneurial environment. These apparent differences in entrepreneurial environment drew us to this international comparison study, which is designed to focus on perceived differences between entrepreneurs and their businesses in Japan and Silicon Valley.

## Methodology

A three-part questionnaire divided into sections on personal difference, corporate difference, and biographical information was designed to measure the perceived entrepreneurial spirit of people (personal difference) and companies (corporate difference). We kept the number of questions to a minimum (100) to obtain maximum responses from real entrepreneurs.

### Measuring Personal Identity by the Personal Difference Index (PDI)

Everyone has two competing desires: to be similar to other people, and to be different from other people. These two conflicting desires are in continual struggle with each other, and eventually reach an equilibrium state. Personal identity is reflected in the difference between one's personal equilibrium and the equilibrium of others. To be able to say, "I am different from other people," brings about strength of identity, and it is the motivation behind creative work in art, science and entrepreneurial activities.

The first part of the questionnaire (Table 10.1) was designed to measure a PDI (one's perceived difference of oneself from other people) by asking entrepreneurs how they think they differ from people who work in large corporations. If strength of identity is the key to the entrepreneurial spirit and can be measured quantitatively, it should then be possible to measure the intensity of entrepreneurial behavior.

### Corporate Identity Measured as Corporate Difference Index (CDI)

Corporations, like people, also have two conflicting desires: to be similar to other corporations, and to be different from other corporations. We assume that these conflicting strategies eventually reach an equilibrium difference, a constant difference between the equilibrium

140  Chapter Ten

## Table 10.1
## Samples of PDI Questions

Section 1-A
How do you respond to the following statement? Circle the applicable number in each case (from 1—strongly agree to 5—strongly disagree).

1. I personally prefer:

   a. to earn a lot of money rather than prove my ability.  1-2-3-4-5

   b. to have the security of working in a stable organization rather than operating my own businesses.  1-2-3-4-5

   c. to tackle a difficult task rather than to carry out my everyday tasks.  1-2-3-4-5

Section 2-A
How do you think the average person in a large corporation would respond to the following statement? Circle the applicable number in each case.

1. Average people in large firms prefer:

   a. to earn a lot of money rather than prove their ability.  1-2-3-4-5

   b. to have the security of working in a stable organization rather than operating their own businesses.  1-2-3-4-5

   c. to tackle a difficult task rather than to carry out their everyday tasks.  1-2-3-4-5

Note: Section 1-A and 2-A have twenty questions each.

---

state of one's company and other companies, measured as a Corporate Difference Index (CDI). This part of the questionnaire (Table 10.2) was developed to measure the respondents' perceived difference of their firm from other firms, especially large corporations.

## Table 10.2
## Samples of CDI Questions

Section 1-B
(1—strongly agree to 5—strongly disagree)

1. The way my firm is run:

    a. it takes a long time to make a decision and put it into action.     1-2-3-4-5

    b. failure does not affect promotion and career progress.     1-2-3-4-5

    c. there are too many reports/meetings.     1-2-3-4-5

Section 2-B
(1—strongly agree to 5—strongly disagree)

1. The way average large firms are run:

    a. it takes a long time to make a decision and put it into action.     1-2-3-4-5

    b. failure does not affect promotion and career progress.     1-2-3-4-5

    c. there are too many reports/meetings.     1-2-3-4-5

Note: Section 1-B and 2-B have twenty questions each.

---

To be able to say, "My company is different from other companies," brings about a strong sense of corporate identity, creating a culture that, in turn, produces innovative activities such as new business development and corporate venturing. If the strength of corporate identity can be measured quantitatively, then it should also be possible to measure the strength of the entrepreneurial environment.

### Biographical Characteristics

This part of the questionnaire included personal and business information about respondents. Twenty questions were asked about age, education, and jobs and job experience, as well as type of business, sales amount, number of years in operation, and number of employees; the answers were correlated with the PDI/CDI data.

### Survey

A simple definition of the entrepreneur was used in this study: An entrepreneur is a person who has founded his or her own enterprises.[2]

*Silicon Valley*

We mailed the questionnaire to 2,475 presidents in Silicon Valley, in October 1988, using the CORPTECH directory. We received 184 responses from founder/co-founders, and 67 from non-founders.

*Japan*

We mailed the questionnaire to 935 Japanese businessmen, in February 1987, and received 305 responses from entrepreneurs, intrapreneurs, and managers, achieving a response rate of over 30%. The questionnaire was completed by 125 (response rate = 35.1%) entrepreneurs who had started their own businesses. This study did not impose any limitations on the age of entrepreneurs or the number of operating years of their ventures.

## Japanese and U.S. High-Tech Entrepreneurs

We used the absolute value of the difference between answers to paired questions in analyzing the PDI/CDI questionnaire because the absolute value of the difference between a pair of answers is assumed to be invariant under various psychological states. This is analogous to the invariant theory of vowels: The absolute difference between two main resonance frequencies determines a specific vowel sound.[3] Thus, taking the absolute value of the difference between perceptions often increases the accuracy of results. For example, this analysis used the absolute difference between the answers to "How do you react to the following statement?" (See Tables 10.1 and 10.2.)

We calculated PDI and CDI by summing the absolute value of the differences between responses given in the relevant sections of the questionnaire and adjusting to a scale from 0 to 100. For example, the PDI/CDI score is 100 when a questionnaire has the maximum possible sum of absolute differences, that is, 80 (twenty pairs of responses), with a maximum of difference of 4 (5 minus 1 for each pair). The minimum possible score is 0 (twenty pairs of responses with a minimum difference of 0 for each pair).

## Background Comparison

Table 10.3 compares the backgrounds of entrepreneurs in Japan and Silicon Valley. As expected, parents' occupations have a great influence on Japanese entrepreneurs, half of whom had parents who were also entrepreneurs. Only one-quarter of the U.S. high-tech entrepreneurs had parents who were entrepreneurs.

Moreover, the education levels of entrepreneurs differ significantly between Japanese and U.S. high-tech entrepreneurs. Approximately 9% of the Japanese entrepreneurs attended graduate school, compared with 57% of the U.S. high-tech entrepreneurs.

Thus, we may hypothesize that in Japan, the family environment is very influential in becoming an entrepreneur. In the United States, family influence is rather small, and education plays a vital role in promoting the entrepreneurial career route, especially in Silicon Valley.

## Correlation Analysis

We computed a correlation matrix of PDI, CDI, and biographical characteristics examined in the study, and found that the following background variables were correlated with PDI or CDI (Table 10.4).

## Table 10.3
## Comparisons of Backgrounds

|  | Japanese entrepreneurs N = 125 | | U.S. High-tech entrepreneurs N = 184 | |
|---|---|---|---|---|
| Background | Mean | SD | Mean | SD |
| No. of years in operation | 16.3 | | 8.9 | |
| No. of employees | 141.5 | | 80.1 | |
| Age (years) | 49.1 | | 47.2 | |
| Age at start-up | 35.0 | | 33.9 | |
| PDI | 41.9 | 13.7 | 38.2 | 10.3 |
| CDI | 36.8 | 11.5 | 36.2 | 9.0 |
| Educational background | (%) | | (%) | |
|   Graduate | 9.1 | | 57.1 | |
|   Undergraduate | 61.2 | | 35.7 | |
|   Technical/certificate | 16.5 | | 1.6 | |
|   High School/matriculation | 10.7 | | 4.9 | |
|   Junior high school | 2.5 | | 0.5 | |
| Occupation of father | (%) | | (%) | |
|   Business owner | 49.2 | | 25.4 | |
|   Large firm/civil servant | 16.9 | | 27.1 | |
|   Small/medium firm (1,000 employees) | 14.5 | | 18.2 | |
|   Specialist | 4.8 | | 10.5 | |
|   Others | 14.5 | | 18.8 | |

*Japanese Sample*

PDIs of Japanese entrepreneurs were correlated with the following backgrounds:

- ○ Number of years of operation (negative)

## Table 10.4
## Correlation Between PDI, CDI and Backgrounds

|  | Japanese entrepreneurs N = 125 |  | U.S. High-tech entrepreneurs N = 184 |  |
|---|---|---|---|---|
| Background | PDI | CDI | PDI | CDI |
| Age | −.10 | −.25* | .07 | .02 |
| No. of years in current firm | −.23* | −.21 |  |  |
| Sales turnover of the company | −.21 | −.29** |  |  |
| No. of employees | −.22* | −.29** | .05 | −.00 |
| Educational level[a] | .09 | .06 | .07 | −.16 |

Significance (one-tailed). *−.01 **−.001
[a]Levels. 1—Junior high school, 2—High school, 3—Technical, 4—Undergraduate, 5—Graduate

- Number of employees (negative)

Their CDIs were correlated with the following:

- Number of employees (negative)
- Sales amount of the company (negative)

This indicates a corporate size effect that depresses PDI and CDI.

We consider these results to be strong evidence that maturation of a Japanese company causes a decrease in CDI. They also suggest that the entrepreneurial spirit is lost with the burden of an increasing number of employees.

### U.S. Sample

Neither the PDIs nor the CDIs of U.S. high-tech entrepreneurs are correlated with any backgrounds. Any type of person can become an entrepreneur in the United States, and entrepreneurial spirit is not lost with the burden of increasing size.

## Boundaries of Entrepreneurial Results

We designated boundaries that separated the results into groups that had distinct properties by plotting PDIs on the horizontal axis and CDIs on the vertical axis, using a scale of 0 to 100. The lowest PDI/CDI ratio (0 points) means that the person/company is exactly the same as others. The highest PDI/CDI (100 points) means that the person/company is maximally different from others.

### Entrepreneurial Line (E-Line)

Figure 10.1a shows the plots of Japanese entrepreneurs whose businesses have been in operation for the last five years. We designated an Entrepreneurial Line (CDI = –0.95PDI + 63), which delimits a boundary that separates the 80% of start-up companies that had been established within the last five years and had high perceived differences from the 20% that had been established in the same time period but had low perceived differences.

Figure 10.1b shows that the E-Line of U.S. high-tech entrepreneurs is the same as that of Japanese entrepreneurs. Thus, entrepreneurial phenomena in the different environments require a constant level of perceived difference.

### Balance Line (B-Line)

Figure 10.2a presents the ID-points of Japanese entrepreneurs and the Balance Line (B-Line), which is the regression line for entrepreneurs (CDI = 1.05 PDI – 7.2). The B-Line marks a balanced combination of PDI and CDI. The ventures that have not passed the five-year start-up yet are widely scattered around the B-Line. A venture falls closer to the B-Line as the venture's years in operation increase, suggesting that ventures close to the B-Line have a higher probability of survival.

## Figure 10.1

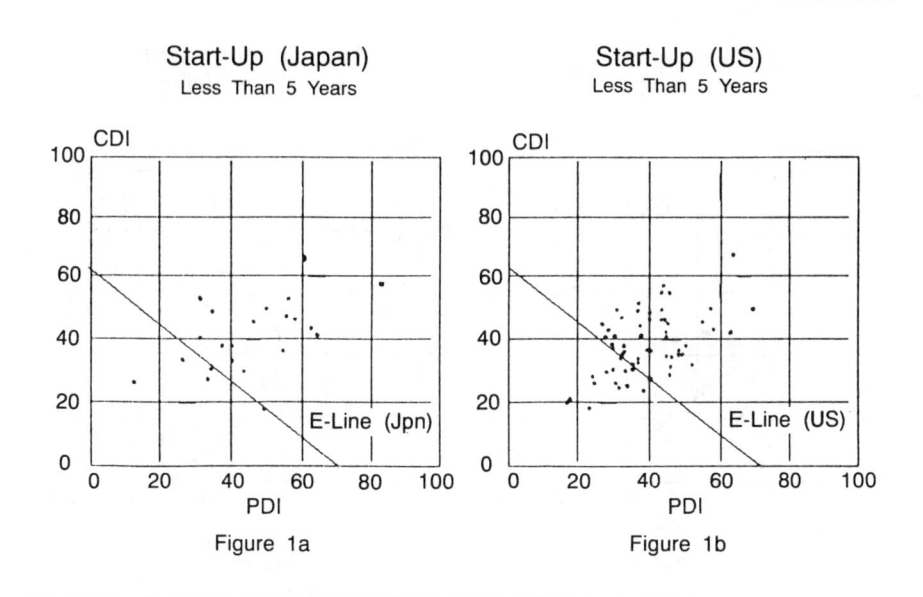

**Start-Up (Japan)**
Less Than 5 Years

**Start-Up (US)**
Less Than 5 Years

Figure 1a

Figure 1b

---

Figure 10.2b shows the ID-points of U.S. high-tech entrepreneurs and their B-Line (CDI = 1.0 PDI −2.0), which is shifted toward the left (negative direction of PDI). The disparity in the entrepreneurial environments of the two countries explains this difference between the B-Lines. In Japan, it is very difficult to become an entrepreneur because of the weak entrepreneurial environment and the emphasis placed on careers other than those in entrepreneurship. As a result, Japanese entrepreneurs require a strong perceived personal difference to overcome the obstacles to entrepreneurship. In contrast, the strong entrepreneurial environment in the United States allows those with less perceived personal difference (negative shift in PDI) to become entrepreneurs.

The comparison of the standard deviation and the EID graphs indicate that U.S. ventures are tightly located around the intersection of the E-Line and the B-Line as well as along the B-Line, indicating again that the U.S. entrepreneurial environment is easier than that in Japan.

## Figure 10.2

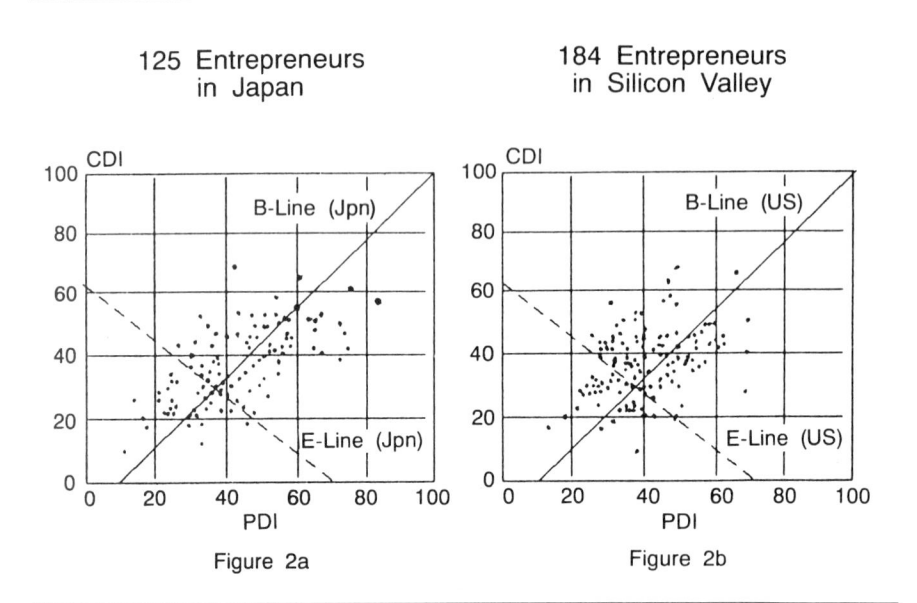

**125 Entrepreneurs in Japan**

Figure 2a

**184 Entrepreneurs in Silicon Valley**

Figure 2b

### Growth and Size

Figure 10.4b shows the scattered ID points of Japanese firms with fewer than 50 employees. Those with more than 50 employees lie along the B-Line and near the intersection of the E-Line and the B-Line, (which we designated the successful venture area), as shown in Figure 10.4a.

U.S. high-tech ventures generally show the same results (Figures 10.3a and 10.3b), with scattered ID points for ventures employing fewer than 50 people. However, U.S. ventures with more than 50 people are more concentrated along the E-Line than Japanese ventures, indicating that the fast-growth firms are located near the successful venture area.

Subsequent interviews with venture capitalists gave us an important insight into these results: Companies that are ready to become public companies tend to be distributed in the successful venture area. These companies have grown in size and have reached the transitional stage from being ventures to becoming medium-sized companies. Entrepreneurs of the slow-growth firms below the successful

## Figure 10.3

**Fast Growth (US)**
Over 50 Persons; Less Than 10 Years

**Slow Growth (US)**
Less Than 50 Persons; Over 10 Years

Figure 3a

Figure 3b

---

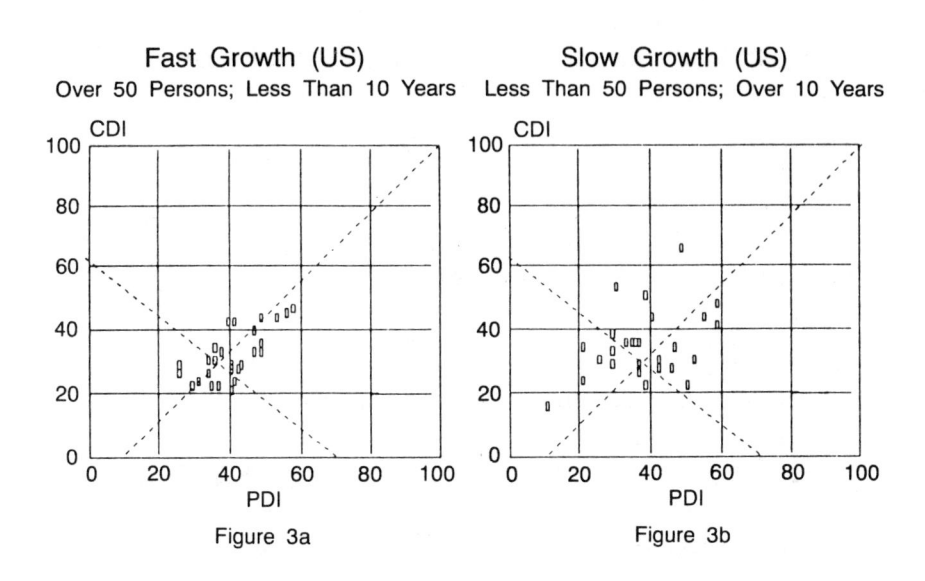

venture area lack entrepreneurial spirit. On the other hand, those above the venture area are *too* different, and are presumably misfits. Because an entrepreneur conducts business through other people, an entrepreneur cannot succeed if he or she is too different. However, he or she cannot be an entrepreneur without some sense of being different from others. In order to be successful, then, entrepreneurs must be different but must not be *too* different.

## Implications

### Comparison of U.S. and Japan

The shared aspects of Japanese and U.S. entrepreneurs are the following:

- Minimum perceived difference of entrepreneurial phenomena

150   Chapter Ten

**Figure 10.4**

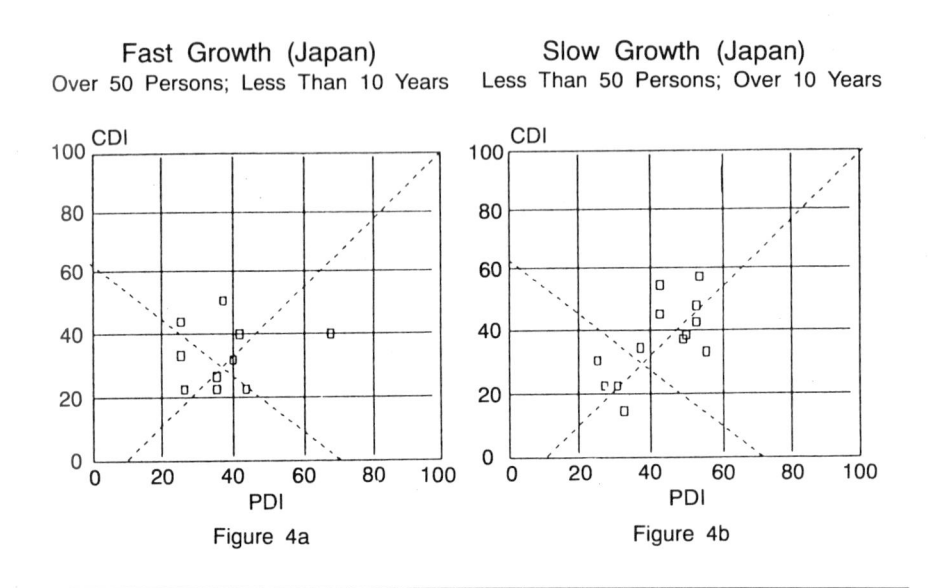

```
        Fast Growth (Japan)              Slow Growth (Japan)
   Over 50 Persons; Less Than 10 Years   Less Than 50 Persons; Over 10 Years
```

Figure 4a                Figure 4b

○ Successful Venture Area: different but not too different

Different aspects of entrepreneurs in U.S. and Japan are these:

○ Background: Occupation of father; education

○ Correlation between PDIs, CDIs and background

○ Shifted B-Line

The same level of motivation is required for entrepreneurial phenomena in Japan and in Silicon Valley, suggesting the existence of a minimum "perceived difference" and an optimized personality for successful entrepreneurs. In Japan, however, a limited number of charismatic people can realize the entrepreneurial phenomenon. Highly educated people rarely leave large corporations in Japan. Combined, these factors cause Japanese ventures to be difficult to start and to grow too slowly. In the United States, any person with an advanced education and the minimum amount of perceived difference can become an entrepreneur.

**Future Research**

First, a study of U.S. middle managers compared with U.S. entrepreneurs and Japanese managers should be done. An expanded and more systematic version of this study could also have practical applications in deciding on new venture investment. The sampling criteria should be expanded. It would be preferable to examine whole management teams rather than just the founders of ventures. Corporation-wide examination of human resources would make it possible to pinpoint entrepreneurial management resources and environments within a corporation. The methods of this study could also be applied to finding researchers who, rather than duplicating the work of others, are pursuing original activities.

# Endnotes

1. Takeru Ohe, Shuji Honjo, and Ian C. MacMillan, Japanese enterpreneurs versus corporate managers: They are different, *Frontiers of Entrepreneurship Research*, 1988, pp. 63–65.

2. Thomas M. Begley and David P. Boyd, Winter 1987. Psychological characteristics associated with performance in entrepreneurial firms and small business, *Journal of Business Venturing*, (1):79–93.

3. Potter, Kopper, Green, 1947. *Visible Speech*, New York, p. 60.

# Chapter Eleven

# Accessing Foreign Venture Capital Sources

## Joel S. Marcus

In ever increasing numbers, U.S. technology companies are finding it necessary to look beyond America's shores to fund their operations, whether by necessity or expediency. All too often, a corporate manager's view is to give away potentially valuable markets that are difficult to understand or penetrate. However, given the current uncertain economic climate, successful managers and their advisors must now move quickly to understand how to access the most "leveraged" corporate partner or foreign venture capital source. At the same time, they must safeguard against giving away their most valuable markets and losing technological "corporate jewels."

Competition for market dominance in technology is fierce. Strategic alliances with foreign companies are fast becoming a way of life in virtually every industry. Increasingly, many American companies have turned to foreign partners for assistance in dealing with intensifying global competition, penetrating foreign markets and shouldering the heavy costs of developing sophisticated new products. The simple and naive act of licensing a foreign company to manufacture and perhaps sell for you, may result in putting another competitor into the marketplace.

To ensure that technology, manufacturing and distribution expertise are shared, American managers should be involved in the overseas venture. Total management control of an overseas venture should never be given away without careful study. An all-too-often undervalued American bargaining chip is the acknowledged strategy of many foreign firms to enter into such corporate alliances to head-off American protectionism.

One successful pattern for this more thoughtful approach to strategic partnering was recently considered by a U.S. venture-backed firm. The U.S. company and its Japanese partner—which was diversifying into the particular market—wish to cooperate in developing that market in Japan, the U.S. and Europe. The firms also wish to co-market present products and future improvements on a worldwide basis. The initial starting point would be the establishment of a 50:50 joint venture in Japan, staffed by both companies.

So what's stopping you? First, you do not speak the language. Second, you do not have a strong familiarity with the culture or the range of potential small, medium and large strategic partners. Third, and most importantly, you do not have the "proper" entree to a targeted group of the most "leveraged" corporate partners/venture capital sources. In this latter respect, who you know is even more important in Japan than in the U.S. Very few U.S. venture capitalists respond to unsolicited business plans sent without a "trusted" referral source. You cannot reasonably expect to locate a very good and reliable corporate partner and/or venture capital source halfway around the globe on your own.

To start this process, it is best to establish a relationship with advisors and experts who have special expertise and substantial experience in such international transactions. Special counsel or special engagement relationships are very common in these circumstances. Knowledge and experience in your industry is also mandatory. You must then utilize these advisors to network you and establish initial contacts with a carefully targeted group of potential partners. This group should even include the spectrum of companies not even in the same line of business, seeking diversification opportunities, to industry leaders who are in your business. Your advisors should be knowledgeable of the many ways in which such choices can be developed at a reasonable cost. Once such companies are identified and an assessment of their strengths and weaknesses developed, you, your advisors and their contacts, can commence the introductory process. Without the "right" introduction, your efforts will generally be fruitless.

A technology company should be very wary of engaging a finder that is too large and where the remuneration is just not significant enough to drive the "location" process and the transaction ahead. For example, a U.S. technology company signed an engagement letter with a large, well-known finder. Nine months later, after virtually no movement in the location process, the technology company was ready to call it quits. This is a frequent occurrence. An intermediary possessing a knowledgeable niche appears to be a more efficient and leverageable resource.

After initial introductions, you should develop as much information about the potential partners as possible regarding its manufacturing, marketing and technological capabilities. Early synergy between the parties is critical. Otherwise, you may be wasting a lot of time and energy. It is also wise to have foreign language capability during each phase of the process—the introductory, negotiation and operational phases. Once you have interviewed the potential partners in depth, you need to rank them according to their capabilities and synergy, as well as from the standpoint of potential future competitiveness. Your advisors should also assist you in this assessment.

Once you have developed a positive response from at least two, and possibly three, target companies, you need to address and assess the relative contributions of your firm and the various targeted companies. As discussed above, a detailed understanding of the potential foreign corporate partners and their strategies must also be developed. This will enable you to crystallize your strategic vision and develop the key deal points. You and your "smart team" must develop this critical focus before you engage in any further exchange of information.

One of the most difficult issues for U.S. technology companies to grapple with is the development of an ability to evaluate the goals and economic parameters of the strategic partnering relationship. You must understand how to leverage your technology with respect to the specific foreign party with which you are dealing.

For example, the receipt of creatively structured fees—to be treated as revenues to the U.S. company—designed to recoup the cost of research and development respecting a particular technology or product, can be negotiated. You should not lose sight of important structuring ideas that avoid the payment of foreign taxes. The unexpected, and unpleasant, requirement to pay foreign taxes will have the effect of substantially reducing your net economic benefit. Many U.S. companies that do not adequately research and address these issues beforehand are caught by surprise and suffer significant negative economic consequences.

Similarly, certain types of payments from foreign entities, e.g., for research and development, are tax-favored. Knowledge of these incentives and your ability to take advantage of such opportunities may inure to your economic benefit during the negotiating process.

You should always be cognizant of seeking to structure the transaction to enhance your profit & loss (P & L). It may also be wise to tie the technology side of the deal to a substantial equity participation in your company by the foreign partner. This type of valuable and much needed venture capital infusion can be easily overlooked. Such a participation may tie the parties together on a longer-term basis and potentially deter destructive downstream competition.

In one significant transaction, after the introductory meeting, the chief executive officer (C.E.O.) of a U.S. technology company took the highly unusual and unfortunate step of faxing a proposed draft of the transaction documents to its purported Pacific Rim corporate partners. No response was ever received from the foreign entity. This is definitely not the way to commence the transaction process.

Before disclosing any revealing, detailed proprietary technological know-how, the undertaking of any validation of the technology or its commercialization potential by the target company, you must first understand the strategic objective of the foreign entity. To facilitate this end, one of the most efficient and effective tactics would be to prepare a one-page executive term sheet of the key business elements of the deal in reasonably general terms. Your advisors and experts can be invaluable in assisting you at this point "from behind the scenes." You want to be specific enough to begin the negotiation process. However, you do not want to be overly specific.

This may have the effect of foreclosing your options or limiting your flexibility downstream. It is also useful to explain in general terms your strategic visions and objectives. Then, and only then, does it make sense to analyze and propose the details of a typical term sheet.

It is commonly stated and acknowledged among technology C.E.O.s that effective and efficient management of corporate partnering relationships presents a difficult challenge and a great drain and strain on management time and resources. The commitment to strategic partnering as an important corporate strategy must be accepted, not only at the top level, but through the management and research levels as well. These relationships could easily falter at the mid-level management and research levels if strong C.E.O. guidance and commitment are not in evidence.

A smart team of top management must be committed to the project. One or more specific "liaison team members" must be designated as the focal point for communication with the corporate partner. A great deal of time, energy and determination must be invested into these relationships. Management must also be perceptive enough to ensure that such relationships maintain their responsiveness to both technological and market changes. Both parties must understand the strategic objective, needs and changing desires of the other.

Underfunded and undermanned technology companies often pursue these relationships when they are in a survival mode. They are not in a position to effectively carry through the dynamic management of the relationship from the introductory process into the operational phase. This can be fatal. On a number of occasions this has led to cancellation of a contract or the withdrawal of strong support and active participation.

Knowledge, trade secrets, confidential process information, manufacturing experience and the like component to be carefully utilized in a corporate partnering transaction, as well as the important technical assistance that accompanies the disclosure or transfer thereof. Similarly, the worth of technical assistance must be fully recognized and not undervalued or discarded.

International corporate partnering/venture capital transactions are inherently fraught with potentially fatal pitfalls. However, such transactions may enable the "smart team" to ensure the survival and growth of the company in the technological business jungle.

*Joel Marcus is a partner in the Los Angeles office of Brobeck, Phleger and Harrison, a West Coast law firm. He specializes in producing U.S.-Japan international technology agreements, and cross-border M & A.*

# Japanese Venture Capital Sources*

*These firms are members of the Venture Enterprise Center located in Tokyo.

**Company:** **Nippon Enterprise Development Corp.**
President: Hiroaki Ueda
Address: JBP Oval Bldg. 3F.
5-52-2, Jinguumae, Shibuya-ku,
Tokyo, 150
Telephone: 03-3797-9560
Established: November 1972

---

Key Contact
Capital Managed / Advised
Type of Financing
Seed
Start-up
First-stage
Second-stage
Leveraged buyout

Minimum Investment
Preferred Investment

Industry Preferences

Comments

## 162  Japanese Venture Capital Sources

**Company:** **Sumigin Finance Co., Ltd.**
President: Masatake Nakano
Address: Sumitomo Shinbashi Bldg. 3F.
1-8-3, Shinbashi, Minato-ku,
Tokyo, 105
Telephone: 03-3573-4661
Established: December 1972

---

Key Contact
Capital Managed/Advised
Type of Financing
Seed
Start-up
First-stage
Second-stage
Leveraged buyout

Minimum Investment
Preferred Investment

Industry Preferences

Comments

**Company:** **Japan Associated Finance Co., Ltd.**
President: Masaki Yoshida
Address: Toshiba Bldg. 10F.
 1-1-1, Shibaura, Minato-ku,
 Tokyo, 105
Telephone: 03-3456-5101
Established: April 1973

---

Key Contact
Capital Managed/Advised
Type of Financing
Seed
Start-up
First-stage
Second-stage
Leveraged buyout

Minimum Investment
Preferred Investment

Industry Preferences

Comments

## 164  Japanese Venture Capital Sources

**Company:** **Yamaichi General Finance Co., Ltd.**
President: Yukio Hitomi
Address: Tsukamotosoyama Bldg. 2F.
4-2-15, Ginza, Chuo-ku,
Tokyo, 104
Telephone: 03-3535-5851
Established: December 1973

---

Key Contact
Capital Managed/Advised
Type of Financing
Seed
Start-up
First-stage
Second-stage
Leveraged buyout

Minimum Investment
Preferred Investment

Industry Preferences

Comments

**Company:** **Central Capital Co., Ltd.**
President: Kanji Terao
Address: Nihonbashi-Tokai Bldg. 5F.
1-7-17, Nihonbashi, Chuo-ku,
Tokyo, 103
Telephone: 03-3273-7721
Established: January 1974

---

Key Contact
Capital Managed/Advised
Type of Financing
Seed
Start-up
First-stage
Second-stage
Leveraged buyout

Minimum Investment
Preferred Investment

Industry Preferences

Comments

## Japanese Venture Capital Sources

**Company:** **Techno-Venture Capital Co., Ltd.**
President: Yaichi Ayukawa
Address: Diamond Plaza Bldg. 3F.
25, Ichibancho, Chlyoda-ku,
Tokyo, 102
Telephone: 03-3262-3131
Established: February 1974

Key Contact
Capital Managed/Advised
Type of Financing
Seed
Start-up
First-stage
Second-stage
Leveraged buyout

Minimum Investment
Preferred Investment

Industry Preferences

Comments

Japanese Venture Capital Sources 167

**Company:** Tokyo Venture Capital Co., Ltd.
President: Hirohisa Matsudaira
Address: Ichikan-Kayabacho Bldg. 3F.
1-6-10, Nihonbashi-kayabacho, Chuo-ku,
Tokyo, 103
Telephone: 03-3662-8961
Established: April 1974

---

Key Contact
Capital Managed/Advised
Type of Financing
Seed
Start-up
First-stage
Second-stage
Leveraged buyout

Minimum Investment
Preferred Investment

Industry Preferences

Comments

168  Japanese Venture Capital Sources

**Company:** Diamond Capital Co., Ltd.
President: Yukihiro Miyake
Address: Kandamitsubishi Bldg. 6F.
3-6-3, Kandakajioho, Chiyoda-ku,
Tokyo, 101
Telephone: 03-3252-4591
Established: August 1974

Key Contact
Capital Managed/Advised
Type of Financing
Seed
Start-up
First-stage
Second-stage
Leveraged buyout

Minimum Investment
Preferred Investment

Industry Preferences

Comments

## Japanese Venture Capital Sources

**Company:** Mitsui Finance Service Co., Ltd.
President: Haruo Inoue
Address: Nishishinbashi-Mitsui Bldg. 2F.
1-24-14, Nishishinbashi, Minato-ku,
Tokyo, 105
Telephone: 03-3502-8005
Established: December 1979

---

Key Contact
Capital Managed/Advised
Type of Financing
Seed
Start-up
First-stage
Second-stage
Leveraged buyout

Minimum Investment
Preferred Investment

Industry Preferences

Comments

## 170 Japanese Venture Capital Sources

**Company:** **Nippon Investment & Finance Co., Ltd.**
President: Takuro Isoda
Address: Shinjuku-Center Bldg. 47F.
1-25-1, Nishishinjuku, Shinjuku-ku,
Tokyo, 163
Telephone: 03-3349-0961
Established: August 1982

---

Key Contact
Capital Managed / Advised
Type of Financing
Seed
Start-up
First-stage
Second-stage
Leveraged buyout

Minimum Investment
Preferred Investment

Industry Preferences

Comments

**Company:** Sanyo Finance Co., Ltd.
President: Yoshihisa Yamaguchi
Address: Landio-Nihonbashi Bldg. 6F.
2-16-13, Nihonbashi, Chuo-ku,
Tokyo, 103
Telephone: 03-3281-3481
Established: August 1982

---

Key Contact
Capital Managed / Advised
Type of Financing
Seed
Start-up
First-stage
Second-stage
Leveraged buyout

Minimum Investment
Preferred Investment

Industry Preferences

Comments

## Japanese Venture Capital Sources

**Company:** Shroder PTV Partners Co., Ltd.
President: Nobuo Matsuki
Address: Jupiter-Uni Bldg. 9F.
2-9-7, Honga, Bunkya-ku,
Tokyo, 113
Telephone: 03-5689-0888
Established: September 1982

---

Key Contact
Capital Managed/Advised
Type of Financing
Seed
Start-up
First-stage
Second-stage
Leveraged buyout

Minimum Investment
Preferred Investment

Industry Preferences

Comments

**Company:** New Japan Finance Co., Ltd.
President: Fumio Nakano
Address: Shinnihonshoken-Nihonbashi Bldg. 3F.
1-17-10, Nihonbashi, Chuo-ku,
Tokyo, 103
Telephone: 03-3277-1860
Established: December 1982

---

Key Contact
Capital Managed/Advised
Type of Financing
Seed
Start-up
First-stage
Second-stage
Leveraged buyout

Minimum Investment
Preferred Investment

Industry Preferences

Comments

## Japanese Venture Capital Sources

**Company:** Wako Finance Co., Ltd.
President: Shigeru Sekihara
Address: Higuchi Bldg. 4F.
1-12-2, Nihonbashi-kayabacho, Chuo-ku, Tokyo, 103
Telephone: 03-3663-4076
Established: December 1982

Key Contact
Capital Managed/Advised
Type of Financing
Seed
Start-up
First-stage
Second-stage
Leveraged buyout

Minimum Investment
Preferred Investment

Industry Preferences

Comments

## Japanese Venture Capital Sources 175

**Company:** **Yamatane Investment Co., Ltd.**
President: Kazuji Mizuno
Address: 2nd SK Bldg. 10F.
2-10-5, Nihonbashi, Chuo-ku,
Tokyo, 103
Telephone: 03-3272-5500
Established: December 1982

---

Key Contact
Capital Managed / Advised
Type of Financing
Seed
Start-up
First-stage
Second-stage
Leveraged buyout

Minimum Investment
Preferred Investment

Industry Preferences

Comments

## 176 Japanese Venture Capital Sources

**Company:** Marusan Finance Co., Ltd.
President: Yoshiji Iwai
Address: 2nd SK Bldg. 3F.
2-10-5, Nihonbashi, Chuo-ku,
Tokyo, 103
Telephone: 03-3274-6766
Established: March 1983

---

Key Contact
Capital Managed / Advised
Type of Financing
Seed
Start-up
First-stage
Second-stage
Leveraged buyout

Minimum Investment
Preferred Investment

Industry Preferences

Comments

Japanese Venture Capital Sources 177

**Company:** Okasan Finance Co., Ltd.
President: Kazuo Kirino
Address: 1-17-6, Nihonbashi, Chuo-ku,
Tokyo, 103

Telephone: 03-3272-8496
Established: April 1983

---

Key Contact
Capital Managed / Advised
Type of Financing
Seed
Start-up
First-stage
Second-stage
Leveraged buyout

Minimum Investment
Preferred Investment

Industry Preferences

Comments

## 178 Japanese Venture Capital Sources

**Company:** **Cosmo General Finance Co., Ltd.**
President: Yukihisa Fujiwara
Address: 2-9-4, Nihonbashi, Chuo-ku,
Tokyo, 103

Telephone: 03-3271-8991
Established: May 1983

---

Key Contact
Capital Managed / Advised
Type of Financing
Seed
Start-up
First-stage
Second-stage
Leveraged buyout

Minimum Investment
Preferred Investment

Industry Preferences

Comments

**Company:** Maruman Finance Co., Ltd.
President: Hiromi Makino
Address: Hirokoji-Daiichiseimei Bldg. 4F.
3-1-1, Sakae, Naka-ku,
Nagoya, 460
Telephone: 052-242-1872
Established: June 1983

---

Key Contact
Capital Managed / Advised
Type of Financing
Seed
Start-up
First-stage
Second-stage
Leveraged buyout

Minimum Investment
Preferred Investment

Industry Preferences

Comments

180  Japanese Venture Capital Sources

**Company:** **Nikko Capital Co., Ltd.**
President: Shigeo Hotta
Address: 1-2-5, Nihonbashi-kayabacho, Chuo-ku,
Toyko, 103

Telephone: 03-3660-2911
Established: July 1983

Key Contact
Capital Managed/Advised
Type of Financing
Seed
Start-up
First-stage
Second-stage
Leveraged buyout

Minimum Investment
Preferred Investment

Industry Preferences

Comments

## Japanese Venture Capital Sources

**Company:** Fujigin Capital Co., Ltd.
President: Shunji Tanefusa
Address: Central Plaza Bldg. 4F.
1-1, Kaguragashi, Shinjuku-ku,
Tokyo, 162
Telephone: 03-3235-0141
Established: July 1983

---

Key Contact
Capital Managed / Advised
Type of Financing
Seed
Start-up
First-stage
Second-stage
Leveraged buyout

Minimum Investment
Preferred Investment

Industry Preferences

Comments

182　Japanese Venture Capital Sources

**Company:**　National Enterprise Co., Ltd.
President:　Tadashi Mukai
Address:　National Security Bldg. 4F.
　　　　　1-5-9, Koraibashi, Chuo-ku,
　　　　　Osaka-shi, Osaka, 541
Telephone:　06-222-0013
Established:　August 1983

---

Key Contact
Capital Managed/Advised
Type of Financing
Seed
Start-up
First-stage
Second-stage
Leveraged buyout

Minimum Investment
Preferred Investment

Industry Preferences

Comments

**Company:** Toyo Finance Co., Ltd.
President: Atsushi Matsumoto
Address: Kasho Bldg. 3F.
2-14-9, Nihonbashi, Chuo-ku,
Tokyo, 103
Telephone: 03-3281-1040
Established: October 1983

---

Key Contact
Capital Managed/Advised
Type of Financing
Seed
Start-up
First-stage
Second-stage
Leveraged buyout

Minimum Investment
Preferred Investment

Industry Preferences

Comments

## 184  Japanese Venture Capital Sources

**Company:** **Orix Capital Corp., Ltd.**
President: Toshio Kawabata
Address: World Trade Center Bldg. 37F.
2-4-1, Hamamatsucho, Minato-ku,
Tokyo, 105
Telephone: 03-3435-4890
Established: October 1983

---

Key Contact
Capital Managed / Advised
Type of Financing
Seed
Start-up
First-stage
Second-stage
Leveraged buyout

Minimum Investment
Preferred Investment

Industry Preferences

Comments

Japanese Venture Capital Sources 185

**Company:** Daiichi Capital Co., Ltd.
President: Ken'ichi Tejima
Address: 1-4-12, Nihonbashi-honcho, Chuo-ku, Tokyo, 103

Telephone: 03-3244-2550
Established: October 1983

---

Key Contact
Capital Managed / Advised
Type of Financing
Seed
Start-up
First-stage
Second-stage
Leveraged buyout

Minimum Investment
Preferred Investment

Industry Preferences

Comments

## 186  Japanese Venture Capital Sources

**Company:** Kankaku Investment Co., Ltd.
President: Hiroshi Momose
Address: Kankaku-Honcho Bldg.
3-3-3, Nihonbashi-honcho, Chuo-ku,
Tokyo, 103
Telephone: 03-3270-6481
Established: February 1984

---

Key Contact
Capital Managed / Advised
Type of Financing
Seed
Start-up
First-stage
Second-stage
Leveraged buyout

Minimum Investment
Preferred Investment

Industry Preferences

Comments

Japanese Venture Capital Sources 187

**Company:** **Yokohama Capital Co., Ltd.**
President: Yasuo Kubo
Address: Yokohama Bank Honbu Annex
6-66-3, Kitanakadori, Naka-ku,
Yokohama-shi, Kanagawa, 231
Telephone: 045-201-2912
Established: March 1984

---

Key Contact
Capital Managed / Advised
Type of Financing
Seed
Start-up
First-stage
Second-stage
Leveraged buyout

Minimum Investment
Preferred Investment

Industry Preferences

Comments

## 188  Japanese Venture Capital Sources

**Company:** Techno Investment Co., Ltd.
President: Takeshi Anraku
Address: Diamond Plaza Bldg. 3F.
25, Ichibancho, Chiyoda-ku,
Tokyo, 102
Telephone: 03-3262-9561
Established: April 1974

Key Contact
Capital Managed/Advised
Type of Financing
Seed
Start-up
First-stage
Second-stage
Leveraged buyout

Minimum Investment
Preferred Investment

Industry Preferences

Comments

**Company:** Ryugin Venture Capital Co., Ltd.
President: Ichira Teruya
Address: Ryukyu Lease Bldg. 6F.
1-7-1, Kumoji, Naha-shi,
Okinawa, 900
Telephone: 098-868-1101
Established: April 1984

Key Contact
Capital Managed / Advised
Type of Financing
Seed
Start-up
First-stage
Second-stage
Leveraged buyout

Minimum Investment
Preferred Investment

Industry Preferences

Comments

# 190 Japanese Venture Capital Sources

**Company:** Chibagin Capital Co., Ltd.
President: Hideo Yamazaki
Address: Nihonkasai Chiba Bldg. 5F.
8-4, Chibaminato, Chiba-shi,
Chiba, 260
Telephone: 0472-48-8822
Established: May 1984

---

Key Contact
Capital Managed/Advised
Type of Financing
Seed
Start-up
First-stage
Second-stage
Leveraged buyout

Minimum Investment
Preferred Investment

Industry Preferences

Comments

**Japanese Venture Capital Sources** 191

**Company:** Takugin Capital Co., Ltd.
President: Yukinobu Ujiie
Address: Sapporo-Tokeidai Bldg.
Nishi 2-1, Kita 1 jo, Chuo-ku, Sapporo-shi,
Hokkaido, 060
Telephone: 011-222-7311
Established: July 1984

---

Key Contact
Capital Managed/Advised
Type of Financing
Seed
Start-up
First-stage
Second-stage
Leveraged buyout

Minimum Investment
Preferred Investment

Industry Preferences

Comments

## 192 Japanese Venture Capital Sources

**Company:** Sanwa Capital Co., Ltd.
President: Yoshikazu Karasawa
Address: Shinjuku L-Tower 8F.
1-6-1, Nishishinjuku, Shinjuku-ku,
Tokyo, 160
Telephone: 03-3340-3000
Established: August 1984

---

Key Contact
Capital Managed/Advised
Type of Financing
Seed
Start-up
First-stage
Second-stage
Leveraged buyout

Minimum Investment
Preferred Investment

Industry Preferences

Comments

**Company:** Shizuoka Capital Co., Ltd.
President: Sachio Kitano
Address: Shizuginkaikan 4F.
1980-4, Kusanagi, Shimizu-shi,
Shizuoka, 424
Telephone: 0543-47-2210
Established: August 1984

---

Key Contact
Capital Managed/Advised
Type of Financing
Seed
Start-up
First-stage
Second-stage
Leveraged buyout

Minimum Investment
Preferred Investment

Industry Preferences

Comments

**194 Japanese Venture Capital Sources**

**Company:** Hachijuni Capital Co., Ltd.
President: Sadatoshi Komanba
Address: 178-8, Okada
Nagano-shi, Nagano, 380

Telephone: 0262-27-6887
Established: September 1984

---

Key Contact
Capital Managed/Advised
Type of Financing
Seed
Start-up
First-stage
Second-stage
Leveraged buyout

Minimum Investment
Preferred Investment

Industry Preferences

Comments

**Company:** Hokuriku Capital Co., Ltd.
President: Yasuo Mitsuda
Address: Toyama Marunouchi Bldg. 4F.
1-8-10, Marunouchi, Toyama-shi,
Toyama, 930
Telephone: 0764-31-2440
Established: January 1985

---

Key Contact
Capital Managed / Advised
Type of Financing
Seed
Start-up
First-stage
Second-stage
Leveraged buyout

Minimum Investment
Preferred Investment

Industry Preferences

Comments

## 196  Japanese Venture Capital Sources

**Company:** **Kyoto Investment & Finance Co., Ltd.**
President: Ikuo Ishida
Address: 700, Yakushimaacho,
Karasumatori Matsubara Agaru,
Shimogyo-ku, Kyoto-shi, Kyoto, 600-91
Telephone: 075-361-5701
Established: June 1985

---

Key Contact
Capital Managed/Advised
Type of Financing
Seed
Start-up
First-stage
Second-stage
Leveraged buyout

Minimum Investment
Preferred Investment

Industry Preferences

Comments

**Company:** Tomin Capital Co., Ltd.
President: Yasuo Hayakawa
Address: 2-3-11, Roppongi, Minato-ku, Tokyo, 106

Telephone: 03-3587-7943
Established: July 1985

---

Key Contact
Capital Managed / Advised
Type of Financing
Seed
Start-up
First-stage
Second-stage
Leveraged buyout

Minimum Investment
Preferred Investment

Industry Preferences

Comments

## Japanese Venture Capital Sources

**Company:** **Iyogin Capital Co., Ltd.**
President: Seizo Tada
Address: 1, Minamihoribatacho, Matsuyama-shi, Ehime, 790

Telephone: 0899-33-8804
Established: August 1985

---

Key Contact
Capital Managed / Advised
Type of Financing
Seed
Start-up
First-stage
Second-stage
Leveraged buyout

Minimum Investment
Preferred Investment

Industry Preferences

Comments

**Company:** **Hyogin Capital Co., Ltd.**
President: Toshio Iseki
Address: Hyogo-Nunobiki Bldg. 3F.
2-5-1, Kanomachi, Chuo-ku, Kobe-shi,
Hyogo, 650
Telephone: 078-261-1212
Established: October 1985

---

Key Contact

Capital Managed/Advised

Type of Financing

Seed

Start-up

First-stage

Second-stage

Leveraged buyout

Minimum Investment

Preferred Investment

Industry Preferences

Comments

## Japanese Venture Capital Sources

**Company:** Kokusai Finance Co., Ltd.
President: Shumhel Amano
Address: 1-3-11, Nihonbashi-honoho, Chuo-ku, Tokyo, 103

Telephone: 03-3242-7031
Established: December 1985

---

Key Contact
Capital Managed/Advised
Type of Financing
Seed
Start-up
First-stage
Second-stage
Leveraged buyout

Minimum Investment
Preferred Investment

Industry Preferences

Comments

**Company:** Daiwa Business Investment Co., Ltd.
President: M. Ekuni
Address: Daiwaginko-Kyutaromachi Bldg. 8F.
2-5-28, Kyutaromachi, Chuo-ku, Osaka-shi,
Osaka, 541
Telephone: 06-243-1990
Established: January 1986

---

Key Contact
Capital Managed/Advised
Type of Financing
Seed
Start-up
First-stage
Second-stage
Leveraged buyout

Minimum Investment
Preferred Investment

Industry Preferences

Comments

## Japanese Venture Capital Sources

**Company:** Himegin Finance Co., Ltd.
President: Masatoshi Takashige
Address: 2-4-7, Katsuyamacho, Matsuyama-shi, Ehime, 790

Telephone: 0899-33-8383
Established: May 1988

---

Key Contact
Capital Managed / Advised
Type of Financing
Seed
Start-up
First-stage
Second-stage
Leveraged buyout

Minimum Investment
Preferred Investment

Industry Preferences

Comments

# Japanese Venture Capital Sources

**Company:** Universal Finance Co., Ltd.
President: Suejira Sato
Address: Daiwa Yaesu Bldg. 4F.
1-2-1, Kyobashi, Chuo-ku,
Tokyo, 104
Telephone: 03-3277-3710
Established: October 1987

---

Key Contact
Capital Managed/Advised
Type of Financing
Seed
Start-up
First-stage
Second-stage
Leveraged buyout

Minimum Investment
Preferred Investment

Industry Preferences

Comments

## 204  Japanese Venture Capital Sources

**Company:** **Kyowa Investment Co., Ltd.**
President: Osamu Kinbara
Address: Nihonbashi-Daiei Bldg. 9F.
1-2-6, Nihonbashi-muromachi,
Chuo-ku, Tokyo, 103
Telephone: 03-3270-3311
Established: March 1988

---

Key Contact
Capital Managed / Advised
Type of Financing
Seed
Start-up
First-stage
Second-stage
Leveraged buyout

Minimum Investment
Preferred Investment

Industry Preferences

Comments

**Company:** Saigin Capital Co., Ltd.
President: Yuzo Handa
Address: Nihonbashi-Daiei Bldg. 9F.
1-2-6, Nihonbashi-muromachi,
Chuo-ku, Tokyo, 103
Telephone: 03-3279-3061
Established: March 1989

---

Key Contact
Capital Managed/Advised
Type of Financing
Seed
Start-up
First-stage
Second-stage
Leveraged buyout

Minimum Investment
Preferred Investment

Industry Preferences

Comments

## 206 Japanese Venture Capital Sources

**Company:** Tokyo General Finance Co., Ltd.
President: Masataka Shinozaki
Address: Nihonbashi-Kakigaracho Tokyu Bldg. 3F.
1-29-1, Nihonbashi-kakigaracho,
Chuo-ku, Tokyo, 103
Telephone: 03-3669-3071
Established: March 1989

---

Key Contact
Capital Managed/Advised
Type of Financing
Seed
Start-up
First-stage
Second-stage
Leveraged buyout

Minimum Investment
Preferred Investment

Industry Preferences

Comments

**Company:** NCB Capital Co., Ltd.
President: Katsuhisa Konno
Address: 1-5-5, Kudan-minami, Chiyoda-ku, Tokyo, 102

Telephone: 03-3261-6061
Established: December 1989

---

Key Contact
Capital Managed/Advised
Type of Financing
Seed
Start-up
First-stage
Second-stage
Leveraged buyout

Minimum Investment
Preferred Investment

Industry Preferences

Comments

## Japanese Venture Capital Sources

**Company:** Kogin Investment Co., Ltd.
President: Akitoshi Furuhata
Address: Akasaka Oji Bldg. 7F.
8-1-22, Akasaka, Minato-ku,
Tokyo, 107
Telephone: 03-3497-5321
Established: April 1990

---

Key Contact
Capital Managed/Advised
Type of Financing
Seed
Start-up
First-stage
Second-stage
Leveraged buyout

Minimum Investment
Preferred Investment

Industry Preferences

Comments

**Company:** Asahi Life Capital Co., Ltd.
President: K. Sonoda
Address: 1-7-3, Nishishinjuku, Shinjuku-ku, Tokyo, 163-91

Telephone: 03-3346-8245
Established: November 1990

---

Key Contact
Capital Managed/Advised
Type of Financing
Seed
Start-up
First-stage
Second-stage
Leveraged buyout

Minimum Investment
Preferred Investment

Industry Preferences

Comments

# International Venture Capital Associations

## The Venture Capital Club of Indonesia

To provide a link between venture capitalists, entrepreneurs, the government and the business community to discuss and exchange information and experiences on venture projects.

To promote the interests of the Venture Capital community.

To serve as an avenue for local entrepreneurs to interact with potential investors.

To organise various activities for its members such as educational programs, discussion groups and talks.

To facilitate relationships with other regional and international venture capital clubs and associations.

For further information please contact

Club Secretariat
c/o PT Binaniagatama Perkasa
Chase Plaza Tower 7th Floor
Jl. Jend. Sudirman Kav. 21
Jakarta, Indonesia
Tel: (62) 570-4335
Fax: (62) 570-3681

## Venture Enterprise Center

The Center for the Development of Research and Development Oriented Enterprises

In Japan, VEC was established as a nonprofit foundation on 1 July 1975. The policies devised and implemented by VEC now provide a major pillar for the promotion and cultivation of venture businesses.

VEC's two main activities are:

- to act as surety on loans from banks to domestic venture businesses;
- to act as an information exchange for venture business

VEC's membership roster of more than 330 companies includes 210 small- and medium-sized ones and 46 leading venture capital companies in Japan.

Address    8th Fl, New Diamond Bldg. 1-4-4 Kasumigaseki, Chiyoda-ku, Toyko 100

Telephone    03-503-3041

Fax    03-503-3046

## The Hong Kong Venture Capital Association Ltd.

The Hong Kong Venture Capital Association provides links between Hong Kong's venture capitalists, entrepreneurs, the Government and the business community.

The Association seeks to promote the interests of the venture capital community. It organizes regular events where members can interact and provides educational programs on venture capital.

For information on joining the Association, please contact:

> The Administrator
> 1407 Wing On Centre
> 111 Connaught Road Central
> Hong Kong

Tel: (852) 541-6986

Fax: (852) 854-3730

## Association Française Investiss. Cap.Risque

c/o 3i s.a. 141, av. Ch. de Gaulle
F-95521 Neuilly Cedex
Tel: 33 1/471.511.00
Fax: 33 1/474.531.24

Chmn: Mr. Michel Biégala

## Venture Capital Association of Australasia

Representing the interests of the venture capital industry in Australia by:

- raising the general awareness of the community as to the value of venture capital
- making submissions to the relevant government authorities on changes of taxation and company law to encourage the growth of the Australian venture capital industry

Membership Fees: A$2000

Membership limited to companies directly involved in the venture capital market.

For further information, please contact:

> Mr. Vanda Gould
> Secretary
> Venture Capital Association of Australasia
> c/o Continental Venture Capital Limited
> Level 20, 56 Pitt Street
> Sydney, New South Wales
> Australia 2000

Tel: (61) 2 251 1868

Fax: (61) 2 251 3840

## Philippine Venture Capital Investment Group

The Philippine Venture Capital Investment Group (PVCIG) is a networking forum for Philippine/International venture capitalists, entrepreneurs, the business community and the government.

The group has a set of core members who meet once a month to:

- discuss business opportunities
- expand their network/contacts
- discuss venture capital and its progress in the Philippines, and
- make new friends

If you wish to attend and/or become a core member of the group please get in touch with:

Edmundo S. Isidro
Chairman
Philippine Venture Capital
Investment Group
Gr. Fl. Torre de Salcedo Building
184 Salcedo St., Legaspi Village
Makati, Metro Manila

Tels: 632-818 5606/632-816 2982

Fax: 632-817 7158

## National Venture Capital Association

1655 North Fort Meyer Drive
Suite 700
Arlington, Virginia 22209
U.S.A.

Tel: (703) 528-4370

Daniel T. Kingsley
Executive Director

## Belgian Venturing Association

c/o Euroventures Benelux H. Henneaulaan, 366 B-1930 Zaventern
Tel: 32/2/725.18.38
Fax: 32/2/721.44.35

Chmn: Mr. Paul Verdurme

## Finnish Venturing Association

c/o SITRA Uudenmaankatu 16-20 B
SF-00120 Helsinki
Tel: 358/0/618.991
Fax: 358/0/645.072

Chmn: Mr. Matts Andersson

## Bundesverband Deutscher KBG's

Karolinger Platz 10-11 D-1000 Berlin 19
Tel: 49 30/302.91.81
Fax: 49 30/302.91.83

Chmn: Dr. Günter Leopold
Exec: Dr. Holger Frommann

## Nederlandse Verenig. Van Participatiemaatschappijen

Prinses Beatrixlaan, 5 Postbus 93093
NL-2509 AB's-Gravenhage
Tel: 31 70/347.06.01
Fax: 31 70/381.95.08

Chmn: Mr. H. W. Hagtglas Versteeg
Exec: Drs. E. A. M. Elbertse

## Irish Venture Capital Association

c/o AIB Venture Capital Bankcentre, PO Box IRL-1128 Ballsbridge, Dublin 4
Tel: 353 1/60.47.33
Fax: 353 1/60.49.83

Chmn and Exec: Mr. Pat Durcan

## Norwegian Venture Capital Association

P.O. Box 1863 - Vika N-0124 Oslo 1

Tel: 47 2/838.955
Fax: 47 2/838.929

Chmn: Mr. Reidar Michaelsen
Exec: Mrs. Gerd Espelid

## Ass. Italiana Fin. Invest. Cap. Rischio

Via Cornaggia, 10 I-20123 Milan
Tel: 39 2/345.25.56
Fax: 39 2/345.25.50

Chmn: Prof. Marco Vitale
Exec: Ms. Anna Gervasoni

## Ass. Portuguese De Capital De Risco

R. Tierno Galvan - Edificio Amoreiras Torre 3-12-P-1200 Lisboa
Tel: 351 1/690.663
Fax: 351 1/690.729

Chmn: Mr. Nuno M. C. De Brito

## International Venture Capital Associations

### AS. Española De Ent. Cap. Riesgo

Castello, 36 - 5A E-28001 Madrid

Tel: 34 1/577.47.59
Fax: 34 1/577.47.53

Chmn: Mr. José Creixell
Exec: Mrs. Dominique Barthel

### British Venture Capital Association

3 Catherine Place GB-London SW1E 6DX
Tel: 44 71/233.52.12
Fax: 44 71/931.05.63

Chmn: Mr. Michael Denny
Exec: Ms. Vicky Mudford

### Svenska Venture Capital Foreningen

c/o SIND Research Department
S-117 86 Stockholm

Tel: 46 8/744.94.75
Fax: 46 8/774.09.80

Chmn: Mr. Staffan Elmgren
Exec: Mr. Helge Herzog

### European Venture Capital Association

Keibergpark Minervastraat 6 - box 6
B-1930 Zavantem
Tel: 32 2/720.60.10
Fax: 32 2/725.30.36

Chmn: Mr. Miguel Zorita Lees
Exec: Mr. William Stevens, Secretary General

### Swiss Venture Capital Association

c/o U. Busslinger P.O. Box 3027
CH-8031 Zürich
Tel: 41 1/273.15.10
Fax: 41 1/273.24.00

Chmn: Mr. Hugo Wyss
Exec: Mr. Ulrich W. Geilinger

# Index

## A

ACMER Sa (Banque Worms), 26, 28
Advent International, 129, 131
Advent Manufacturing/Marketing International (AMI), xx, 133
Alan Patricof Associates
 APA Excelsior III, 10
 Pan-European fund, 11
American Ventures, Inc., xiii
Anti-Trust Act, 64
Ardent Computer Corporation, 10
Article 154, 75
Article 200, 76
ASEAN, 65, 67, 133

Asian stock market performance, 71

## B

Balance Line (B-Line), 146–148, 150
Bank for International Settlements, 77–78
Bank of Japan, 73–74
Bank of Tokyo, 129
Barclays PLC, 73
Benchaa bijinesu (venture business), xi
Black Monday. *See* October crash
Bridge financing, xiv
Build, Operate and Transfer (BOT), 69
Business Expansion Scheme, 2

219

## C

Capital gains tax, 123
Center for the Development of Research and Development OrientedEnterprises in Japanese, 32
Certified public accountants, 75
Commercial Code of Japan, 110
Computer-aided engineering software, 10
Condominium prices, 72
Continental Bank, 7
Continental Capital Markets, 7
Convertible bonds, 75, 100
Corporate difference index (CDI), 138–151
  entrepreneurial comparison, 142–146
  implications, 149–151
  methodology, 139–142
  results, 146–149
Corporate venture capital (CVC)
  Japanese, 116
  logic, 114–117
  opportunities and challenges, 113–119
Cross-border fund, 11

## D

Dai-ichi Kangyo Bank, xi, 73, 78
Dai-ichi Mutual Life, xxi
Daiwa Bank, 7
Daiwa Securities, 8
  Gilliam/Daiwa Life Science Fund, 8
Deutsche Bank, 73
Dominion Ventures (Dominion Fund II), 7

## E

Employee Stock Ownership Plan (ESOP), 109
Entrepreneurial Line (E-Line), 146-148
Entrepreneurism
  VEC's definition, 35
  Japanese/American differences, 137–151
Eurodollar market, 73
European Community, 3
European Venture Capital Association, 1
Euroventures BV, 4, 28

## F

Fair Trade Commission, 41
Federal Trade Commission (FTC), 126
Foreign
  capital, 153–158
  companies, 107–109, 114–119, 129–136
Fuji Bank, 73, 78
Fundraising, 32

## G

G-5 meeting, xii
GIMV, 26, 28, 64
Globalization, ix

# Index

Guarantee enterprises, 36–38

## H

Hambro Bank, 6
Hambro International, 6, 116
  Equity Partners III (HIP III), 6
High-Tech Companies
  capital and debt source, 71–83
Hiroshima Stock Exchange, xiii
Honda, 124, 134
Hong Kong Stock Exchange, 71

## I

IEP. *See* Investment
Industrial Bank of Japan, 18, 73
Initial Public Offering
  Business, 18–19
  capital markets, 86
  foreign IPO comparison, 106–109
  France, 4
  IPO, xiii–xiv, xvi–xvii, xix–xx, 3, 128
  Japan issuers, 102–103
  new issuers, 105
  OTC comparison, 102–103
  outlook, 105
  Techno-Venture, 130
  underwriters' role, 101–102
  United Kingdom, 4
  venture capital-backed, 61, 125
Insider trading, 75–76
International
  Bioscience Fund-Japan, 7–8
  business strategies, 61–70, 113–119
  network, 121–136
Internationalization, 1–11
Investment
  exiting, 69
  Italy, 25–29
  OTC market, 103–105
  problems in Japan, 126–127
  special interest investors, 109
  United States, 13–24
Investment Enterprise Partnership (IEP), 19–20

## J

Japan
  economic developments, 72–75
  entrepreneurial differences, 137–151
  insider trading, 75–76
  over-the-counter market (OTC), 85–111
  securities companies, 74
  securities regulations, 75–76
  source of capital and debt, 71–83
  United States partnership, 121–136
  venture capital, 123–129
Japan Associated Finance Company (JAFCO), x–xi, 13–14
  *See* Investment
  American Ventures, Inc., xiii
  investment process, 15–16
  IPO business, 19
  Japanese investments, 17

222  Index

Japan Securities Dealers Automatic Quotation System JASDAQ, 19, 88, 100–101
 Preference of Private Equity Investment (U.S.), 16
 Selected U.S. investments, 16
 SOFIPA, 25, 27–28
Japanese
 Construction Ministry, 72
 corporate venture capital, 113–119
 Securities and Exchange Law, 75
  illegal parking, 76
  jay-walking, 76
Japan Securities Dealers Association (JSDA), 87, 110
JASDAQ. *See* Japan Associated

## K

Kenkyuu Kaihatsugata Kigyoo Shinkoo Shitsu, x–xi
Kliener Perkins Caufield & Byers, 64, 129
Korea Kuwait Banking Corporation (KKBC), 67
Kyoto Enterprise Development (KED), xi

## L

Leveraged Buyout (LBO), 15, 18
Living dead company, 16
London Stock Exchange, 2
Long-Term Credit Bank of Japan, xi, 129

## M

Management buyout (MBO), 15, 123
Matsushita, 116, 124
Matuschka Venture Partners, 7–8
Mediocredito, 25–26
Mergers and acquisitions (M&A), 125, 131–132
Ministry of Finance, xii, 74–75, 87, 101, 110
 Securities Bureau, 76
Ministry of International Trade and Industry (MITI)
 *See* venture business
 Center for the Development of Research and Development Oriented Enterprises in Japan, 32
 Techno-Venture, 133
 Venture Enterprise Center (VEC), x, xiii, xv, 32
  enterprise growth, 36–38
  entrepreneurism definition, 35
  guarantee enterprises, 36–38
  loan objectives, 33
  venture capital companies, 41–59
Mitsubishi Corporation, 6–7, 115–116
Mitsui, 115-117, 129
 Trading Company, 6
 USA, 7
Moody's Investor Service, 77
Morgan, J.P., 73

## N

National Association of Security Dealers Automated Quotation NASDAQ, xiii
  OTC comparison, 97–100
  venture-backed firms, 125
NEA. *See* Nippon
NED. *See* Nippon
Network (European-Japanese), 25–29
New Enterprise Associates (NEA), 66–67
Newly Industrialized Economies (NIE), 66–67
Nightingale Market, 2
Nikkei Index. *See* Tokyo Stock Exchange
Nikko Securities, 101
Nippon, 117, 129–130, 132
  Avionics, 106, 108
  Enterprise Development (NED), xi, 67
    evaluation, 64
    international business strategies, 61–70
    internationalization, 63
    investments, 63–64
    NED Delaware Co., 66
    NED Hong Kong Company, 66
  Investment Finance, 4
  Life Insurance, 14, 21
  Steel, 127
Nissan, 117, 134
Nomura Securities, xi, 14, 101
  International, 7

## O

October crash of 1987, 5, 127
Oil crisis, 14, 32
Onset Enterprise Associates (OEA), 66
Orien Venture Capital Fund, 5, 117
Orien II, 6
Osaka Stock Exchange, xiii
Over-the-counter market (OTC), xx, 18, 61
  *See* Nightingale, Tokyo
  foreign IPO comparison, 106–109
  improvements, 87–88
  investors, 103–105
  IPO comparison, 102–103
  Japan, 85–111
  listings, 19
  NASDAQ comparison, 97–100
  new issuers, 105
  outlook, 105
  pricing, 100–101
  procedures, 109–110
  Recruit Cosmo, 127
  registration criteria, 87, 109–110
  stock exchange comparison, 88–97
  trend, 86–87
  venture-backed firms, 125
  venture capital expansion, 4
  volume increase, 99

## P

Partnership funds, xvii

Japanese venture capital, 13–14
Perestroika, 127
Personal Difference Index (PDI), 138–151
  entrepreneurial comparison, 142–146
  implications, 149–151
  methodology, 139–142
  results, 146–149
Price Book Ratio (PBR), 93, 97
Price/earnings ratio (P/E)(PER), xvi, 18, 93, 106
Pricing
  public auction pricing system, 87
Primer Fund. *See* Techno-Venture
Profit and Loss (P&L), 156
Promotion, 31–59

## Q

QUICK, 87–88, 100

## R

Recruit Cosmo, 127
Registration Application, 110
Research and development (R&D), 32–33, 36–40, 115, 118, 129–133
Return on investment (ROI), xvi–xvii, 118
Risk taking, 127

## S

Sanwa Bank, 14, 21
Seconde Marche, 2
Securities and Exchange Commission (SEC), 76
Securities and Exchange Law (of Japan), 110
Securities regulations, 75
Security Registration, 110
Seed/early stage investment, xviii
Silicon Valley, 67, 122
  entrepreneurs, 137–151
Small Business Investment Company, 39
Societa Finanziaria di Participazione S.p.A SOFIPA, 25–29
Sony, 116, 124, 134
Special interest investors, 109
Start-ups, xi, 18
Stock exchange
  *See* individual listings
  OTC comparison, 88–97
Strategy (international business), 61–70, 118
Subchapter S, xiv
Sumitomo Corporation, 4

## T

TA Associates, 64, 129
Taiwan stock market, 71

# Index

Technology transfer, 121–136
  purchase, 124
Techno-Venture, xix–xx, 15, 117, 127
  MITI, 133
  phases, 130
  Primer Fund, 132
  USA Limited Partnership, 132
  U.S./Japanese role, 129–136
Thatcher, Margaret, 2
Tier 2 Capital, 77
Tokyo Over-the-counter (OTC), xii–xiii, 18
Tokyo Stock Exchange (TSE)
  capital relation, 73
  equity, 71–72
  Finance Ministry, 75
  First Section, xiii, 88, 132
  guideline ratios, 77
  Nikkei Average, 97
  Nikkei Index, 72, 135
  Nikkei 225 Index, 93
  rise, 128
  Second Section, xii–xiii, 86, 88, 93, 97, 103, 106, 111
  volatility, 77
Toshiba, 116, 131
Toyota, 116, 134

## U

Underwriter
  Agreements, 101
  role in the IPO market, 101–102

Unlisted Securities Market (USM), 2
United States
  Japanese partnership, 121–136
  Productivity Study Commission, 134

## V

Venture business (benchaa bijinesu)(VB), xi
  characteristics, 32
  definition, xi, 31
  effects, 33
  promotion, 31–59
Venture Business Enterprises, 33–34
Venture capital
  accessing, 153–158
  activities, 38–41
  boom, 14
  companies, 41–59
  Japanese in America, 13–24
  logic, 114–115
  NED, 67
  opportunities and challenges, 113–119
  sources, 153–158
  United States, 66, 123–129
Venture capitalist (VC), x–xii
  development analysis, xvii–xxi
  expertise, 126
  NEA, 66
  Techno-Venture, 131

Venture Enterprise Center
 (VEC). *See* Ministry
Vietnam war, 123

**W**

World War II, 124–125

**Y**

Yamaichi
 Securities, xv, xvii, 101
 services, xv–xvii
YKK, 124, 134